The ABCs of Gene Cloning

Dominic W.S. Wong

The ABCs of Gene Cloning

Second Edition

Springer

LLC

Library of Congress Control Number: 2005931435

ISBN-10: 0-387-28663-2 e-ISBN: 0-387-28679-9
ISBN-13: 978-0387-28663-1

First edition © 1997 by Chapman & Hall

Wong: The ABCs of Gene Cloning

First Indian Reprint 2008

ISBN 978-81-8489-014-3

This edition is manufactured in India for sale in India, Pakistan, Bangladesh, Nepal and Sri Lanka and any other country as authorized by the publisher.

This edition is published by Springer (India) Private Limited,
A part of Springer Science+Business Media, Registered Office: 906-907,
Akash Deep Building, Barakhamba Road, New Delhi – 110 001, India.

Printed in India by Rajkamal Electric Press, Delhi.

Contents

Preface to the second edition

In the nine years since the First Edition, my contention remains that an effective approach to understand the subject of gene cloning is by learning the "vocabulary" and the "language". This book emphasizes the nuts and bolts on just how to do that - reading and speaking the language of gene cloning. It shows the readers how to distinguish between a gene and a DNA, to read and write a gene sequence, to talk intelligently about cloning, to read science news and to enjoy seminars with some degree of comprehension.

On the whole, the second edition is not any more advanced than the first, with the intent of keeping the book concise and not burdening the readers with unwarranted details. Nevertheless, changes were needed and new materials were incorporated in the revision. Part I has a new chapter to provide a tutorial on reading both prokaryotic and eukaryotic gene sequences. Part II consists of several additions, updating on new techniques and cloning vectors. The topics in Part III have been rearranged in separate sections - Part III now focuses on applications of gene cloning in agriculture, and Part IV is devoted entirely to applications in medicine. Chapters on gene therapy, gene targeting, and DNA typing have been thoroughly revised. Additional coverage is included on animal cloning and human genome sequencing. The heavy activity in rewriting and expanding Part IV reflects the rapid progress in the technology and the increased impact of gene cloning.

I enjoyed writing and revising this book with deep satisfaction. It has been an inspiring experience to witness the remarkable development in the field of gene cloning and the tireless dedication of thousands of scientists in making genes tick.

Preface to the first edition

Gene cloning has become a fast growing field with a wide-ranging impact on every facet of our lives. The subject of gene cloning could be intimidating to the novice with little formal training in biology. This book is not intended to give an elementary treatment of recombinant DNA technology, as there are already a number of books in this category. The objective of writing this book is to provide a genuine introduction in gene cloning for interested readers with no prior knowledge in this area to learn the vocabulary and acquire some proficiency in reading and speaking the "language".

In the process of writing this book, the author was continuously confronted with how to present the language of a complex field in a simple and accessible manner. I have chosen to devote Part I of this book to outlining some basic concepts of biology in a straightforward and accessible manner. My intention is to highlight only the essentials that are most relevant to understanding gene cloning. For those who want to pursue a thorough review of genetics or molecular biology, there are many excellent references available. Part II of the book describes cloning techniques and approaches used in microbial, plant, as well as mammalian systems. I believe that a discussion beyond microbes is a prerequisite to a better comprehension of the language and the practical uses of gene cloning. Part III describes selected applications in agriculture and food science, and in medicine and related areas. I have taken the approach to first introduce the background information for each application, followed by an example of cloning strategies published in the literature. The inclusion of publications is an efficient way to demonstrate how gene cloning is conducted, and relate it to the concepts developed in Parts I and II. Moreover, it enables the readers to "see" the coherent theme underlining the principles and techniques of gene cloning. Consistent with its introductory nature, the text is extensively illustrated and the contents are developed in a logical sequence. Each chapter is supplemented with a list of review questions as a study-aid.

I hope that this book will succeed in conveying not only the wonderful language of gene cloning, but also a sense of relevance of this science in our everyday lives. Finally, I acknowledge the contributions of my teachers and colleagues, especially Dr. Carl A. Batt and Dr. Robert E. Feeney, to my persisting interest of biological molecules and processes. Special thanks are due to Dr. Eleanor S. Reimer who has been very supportive in making this book a reality.

Part One

Fundamentals of Genetic Processes

CHAPTER 1

INTRODUCTORY CONCEPTS

The building blocks of all forms of life are cells. Simple organisms such as bacteria exist as single cells. Plants and animals are composed of many cell types, each organized into tissues and organs of specific functions. The determinants of genetic traits of living organisms are contained within the nucleus of each cell, in the form of a type of nucleic acids, called deoxyribonucleic acid (DNA). The genetic information in DNA is used for the synthesis of proteins unique to a cell. The ability of cells to express information coded by DNA in the form of protein molecules is achieved by a two-stage process of transcription and translation.

$$\text{DNA} \xrightarrow{\text{Transcription}} \xrightarrow{\text{Translation}} \text{Protein}$$

1.1 What is DNA and What is a Gene?

A DNA molecule contains numerous discrete pieces of information, each coding for the structure of a particular protein. Each piece of the information that specifies a protein corresponds to only a very small segment of the DNA molecule. Bacteriophage λ, a virus that infects bacteria, contains all its 60 genes in a single DNA molecule. In humans, there are ~31,000 genes organized in 46 chromosomes, complex structures of DNA molecules associated with proteins.

When, how, and where the synthesis of each protein occurs is precisely controlled. Biological systems are optimized for efficiency; proteins are made only when needed. This means that transcription and translation of a gene in the production of a protein are highly regulated by a number of control elements, many of which are themselves proteins. These regulatory proteins are in turn coded by a set of genes.

It is therefore more appropriate to define a gene as a functional unit. A gene is a combination of DNA segments that contain all the information necessary for its expression, leading to the formation of a protein. A gene defined in this context would include (1) the structural gene sequence that encodes the protein, and (2) sequences that are involved in the regulatory function of the process.

1.2 What is Gene Cloning?

Gene cloning is the process of introducing a foreign DNA (or gene) into a host (bacterial, plant, or animal) cell. In order to accomplish this, the gene is usually inserted into a vector (a small piece of DNA) to form a recombinant DNA molecule. The vector acts as a vehicle for introducing the gene into the host cell and for directing the proper replication (DNA -> DNA) and expression (DNA -> protein) of the gene (Fig. 1.1).

The process by which the gene-containing vector is introduced into a host cell is called "transformation". The host cell now harboring the foreign gene is a "transformed" cell.

The host cell carrying the gene-containing vector produces progeny all of which contain the inserted gene. These identical cells are called "clones".

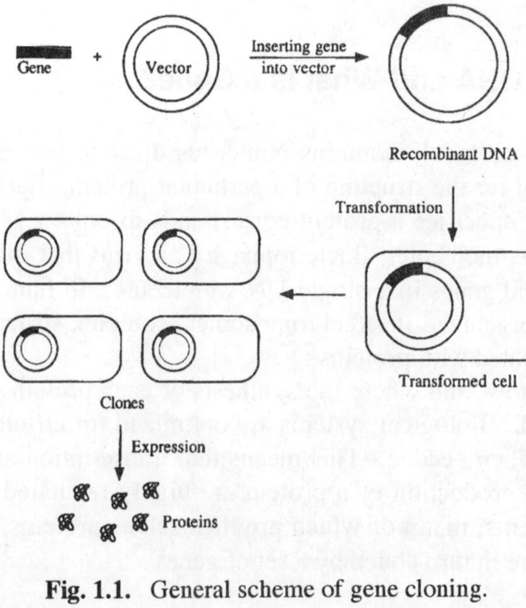

Fig. 1.1. General scheme of gene cloning.

In the transformed host cell and its clones, the inserted gene is transcribed and translated into proteins. The gene is therefore "expressed", with the gene product being a protein. The process is called "expression".

1.3 Cell Organization

Let us focus the attention for a moment on the organization and the general structural features of a cell, knowledge of which is required for commanding the language of gene cloning. Cells exist in one of two distinct types of arrangements (Fig. 1.2). In a simple cell type, there are no separate compartments for genetic materials and other internal structures.

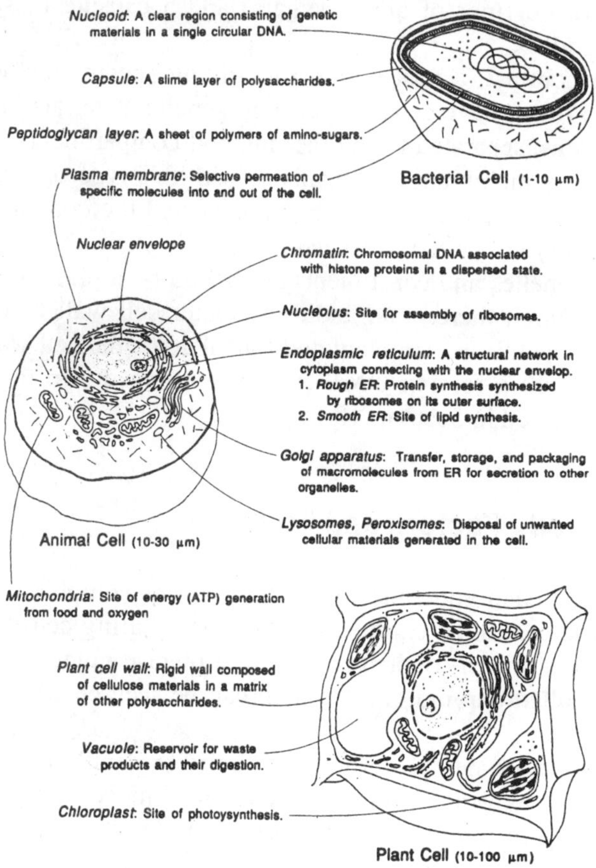

Fig. 1.2. Drawing of cells showing details of organelles.

Organisms with this type of cellular organization are referred to as prokaryotes. The genetic materials of prokaryotes, such as bacteria, are present in a single circular DNA in a clear region called nucleoid that can be observed microscopically. Some bacteria also contain small circular DNA molecules called plasmids. (Plasmids are the DNA used to construct vectors in gene cloning. See Section 9.1) The rest of the cell interior is called the cytoplasm and contains numerous minute spherical structures called ribosomes - the sites for protein synthesis. (Defined structures like ribosomes, are called organelles.) The rest (fluid portion) of the cytoplasm is the cytosol, a solution of chemical constituents that maintain various functions of the cell. All the intracellular materials are enclosed by a plasma membrane, a bilayer of phospholipids in which various proteins are embedded. In addition, some bacterial cells contain an outer layer of peptidoglycan (a polymer of amino-sugars) and a capsule (a slimy layer of polysaccharides).

In contrast, a vast majority of living species including animals, plants, and fungi, have cells that contain genetic materials in a membrane-bound nucleus, separated from other internal compartments which are also surrounded by membranes. Organisms with this type of cell organization are referred to as eukaryotes. The number and the complexity of organelles in eukaryotic cells far exceed those in bacteria (Fig 1.2). In animal cells all organelles and constituents are bound by a plasma membrane. In plants and fungi, there is an additional outer cell wall that is comprised primarily of cellulose. (In plant and fungal cells, the cell wall needs to be removed before a foreign DNA can be introduced into the cell in some cases as described in Sections 10.1 and 10.2.)

1.4 Heredity Factors and Traits

In an eukaryotic nucleus, DNA exists as complex with proteins to form a structure called chromatin (Fig. 1.3). During cell division, the fibrous-like chromatin condenses to form a precise number of well-defined structures called chromosomes, which can be seen clearly under a microscope.

Chromosomes are grouped in pairs by similarities in shape and length as well as genetic composition. The number of chromosome pairs varies among different species. For example, carrots have 9 pairs of chromosomes, humans have 23 pairs, and so on. The two similar chromosomes in a pair are described as homologous, containing genetic materials that control the same inherited traits. If a heredity factor (gene) that determines a specific inherited trait is located in one chromosome, it is also found at the same location (locus) on the homologous chromosome. The two copies of a gene that are found in the same loci in a homologous chromosome pair are determinants of the same hereditary trait, but may exist in various forms (alleles). In simple terms, dominant and recessive alleles exist for each gene.

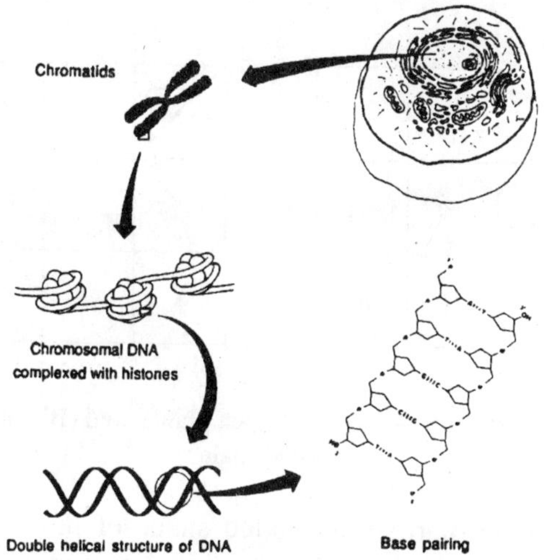

Fig. 1.3. Molecular structure of cellular chromosomes.

In a homologous chromosome pair, the two copies of a gene can exist in three types of combinations: 2 dominant alleles, 1 dominant and 1 recessive, or 2 recessives. Dominant alleles are designated by capital letters, and recessive alleles by the same letter but in lower case. For example, the shape of a pea seed is determined by the presence of the *R* gene. The dominant form of the gene is "*R*", and the recessive form of the gene is designated as "*r*". The homologous combination of the alleles can be one of the following: (1) *RR* (both dominant), (2) *Rr* (one dominant, one recessive) or (3) *rr* (both recessive). This genetic makeup of a heredity factor is called the genotype. A dominant allele is the form of a gene that is always expressed, while a recessive allele is suppressed in the presence of a dominant allele. Hence, in the case of the genotypes *RR* and *Rr*, the pea seeds acquire a round shape, and a genotype of *rr* will give a wrinkled seed. The observed appearance from the expression of a genotype is called its phenotype.

In our example, a pea plant with a genotype of *RR* or *Rr* has a phenotype of round shape seeds. When two alleles of a gene are the same (such as *RR* or *rr*), they are called homozygous (dominant or recessive). If the two alleles are different (such as *Rr*), they are heterozygous. The genotypes and phenotypes of the offspring from breeding between, for example, two pea plants having genotypes of *Rr* (heterozygous) and *rr* (homozygous recessive), can be tracked by the use of a Punnett square (Fig. 1.4A). The offspring in the first generation will have genotypes of *Rr* and *rr* in a 1:1 ratio, and phenotypes of round seed and wrinkled seed, respectively.

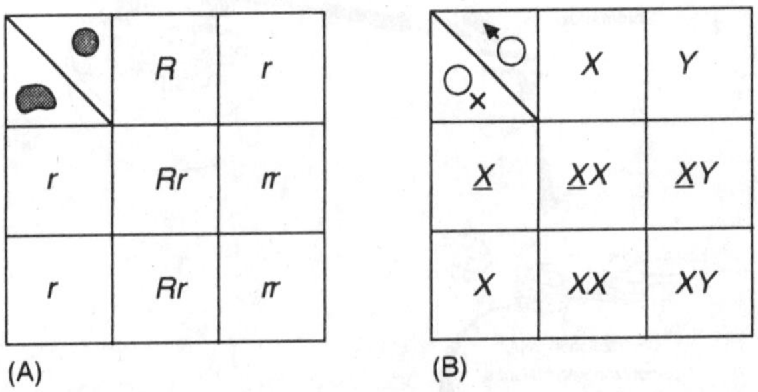

Fig. 1.4. Cross between (A) Rr and rr pea plants, and (B) carrier female and normal male.

The example of round/wrinkled shape of pea seeds is typical of one gene controlling a single trait. The situation is more complex in most cases, because many traits are determined by polygenes. Eye color, for example, is controlled by the presence of several genes. In some cases, a gene may exist in more than two allelic forms. Human ABO blood types are controlled by a gene with 3 alleles - I^A and I^B are codominant, and I^o is recessive. Additional variations are introduced by a phenomenon called crossing over (or recombination) in which a genetic segment of one chromosome is exchanged with the corresponding segment of the homologous chromosome during meiosis (a cell division process, see Sections 1.5 and 17.1).

A further complication arises from sex-linked traits. Humans have 23 pairs of chromosomes. Chromosome pairs 1 to 22 are homologous pairs, and the last pair contains sex chromosomes. Male has XY pair and female has XX chromosomes. The genes carried by the Y chromosome dictate the development of a male; the lack of the Y chromosome results in a female. A sex-linked gene is a gene located on a sex chromosome. Most known human sex-linked genes are located on the X chromosome, and thus are referred to as X-linked. An example of a sex-linked trait is color blindness, which is caused by a recessive allele on the X chromosome (Fig. 1.4B). If a carrier female is married to a normal male, the children will have the following genotypes and phenotype- Sons: $\underline{X}Y$ (color blind) and *XY* (normal), and daughters: $\underline{X}X$ (normal, carrier) and *XX* (normal, non-carrier)

1.5 Mitosis and Meiosis

The presence of homologous chromosome pairs is the result of sexual reproduction. One member of each chromosome pair is inherited from each parent. In human and other higher organisms, autosomal cells (all cells except germ cells, sperms and eggs) contain a complete set of homologous chromosomes, one of each pair from one parent. These cells are called diploid cells ($2n$). Germ cells contain only one homolog of each chromosome pair, and are referred to as haploid (n).

A fundamental characteristic of cells is their ability to reproduce themselves by cell division - a process of duplication in which two new (daughter) cells arise from the division of an existing (parent) cell. Bacterial cells employ cell division as a means of asexual reproduction, producing daughter cells by binary fission. The chromosome in a parent cell is duplicated, and separated so that each of the two daughter cells acquires the same chromosome as the parent cell.

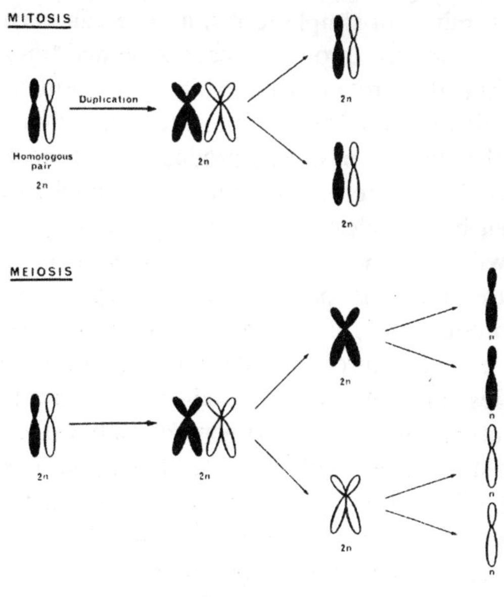

Fig. 1.5. Schematic comparison between mitosis and meiosis.

For eukaryotic cells, the process is not as straightforward. Two types of cell division, mitosis and meiosis, are identified. In mitosis, each chromosome is copied into duplicates (called chromatids) that are separate and partitioned into two daughter cells. Therefore, each of the two daughter cells receives an exact copy of the genetic information possessed by the parent cell (Fig. 1.5). Mitosis permits new cells to replace old cells, a process essential for growth and maintenance. In meiosis, the two chromatids of each chromosome stay attached, and the chromosome pairs are separated instead, resulting in each daughter cell carrying half of the number of chromosomes of the parent cell (Fig. 1.5). Note that at this stage, each chromosome in the daughter cells consists of 2 chromatids. In a second step of division, the chromatids split, resulting in 4 daughter cells each containing a haploid number of chromosomes, i.e. only one member of each homologous chromosome pair. Meiosis is the process by which germ cells are produced. After fertilization of an egg with a sperm, the embryo has complete pairs of homologous chromosomes.

1.6 Relating Genes to Inherited Traits

We will now examine how the preceding discussion of dominance and recessiveness, and genotypes and phenotypes, relate to how genes determine inherited traits. In simple terms, a gene can exist in a functional form, so that it is expressed through transcription and translation to yield a gene product (a specific protein) that exhibits its normal function. However, a gene can also be non-functional due to a mutation, for example, resulting in either the absence of a gene product, or a gene product that does not function properly. Therefore, a homozygous dominant genotype, such as *AA*, means that both alleles in the chromosome pair are functional. A genotype of *Aa* will still have one functional copy of the gene that permits the synthesis of the functional protein. A homozygous recessive (*aa*) individual does not produce the gene product (or produce a nonfunctional gene product). Since a gene controls an inherited trait via the synthesis of the protein it specifies, the presence or absence of a gene product due to a functional or nonfunctional (mutated) gene directly affect a particular inherited characteristic. Genes with multiple alleles can be explained by the difference in the efficiencies of the functions of the gene products. Another explanation is that one copy of the gene produces a lower amount of the gene product than the corresponding normal (functional) gene.

A simple example can be drawn from the genetic disorder of obesity in mice. Obese (*ob*) is an autosomal recessive mutation on mouse chromosome 6. The normal gene encodes the Ob protein which functions in a signal pathway for the body to adjust its energy metabolism and fat accumulation (see Section 17.4). Mice carrying 2 mutant copies (*ob/ob*) of the gene develop progressive obesity with increased efficiency in metabolism (i.e. increase weight gain per calorie intake). Mice with *ob/ob* genotype apparently do not produce the gene product (Ob protein), because both copies of the *ob* gene are nonfunctional.

1.7 Why Gene Cloning?

The general objective of gene cloning is to manipulate protein synthesis. There are several reasons why we want to do this.

(1) To produce a protein in large quantity. Large-scale production of therapeutic proteins has been a primary locus of biotechnology. Many proteins of potential therapeutic values are often found in minute amounts in biological systems. It is not economically feasible to purify these proteins from their natural sources. To circumvent this, the gene of a desirable protein is inserted into a suitable host system that can efficiently produce the protein in large quantities. Examples of pharmaceuticals of this type include human insulin, human growth hormone, interferon, hepatitis B vaccine, tissue plasminogen activator, interleukin-2, and erythropoietin. Another area of great interest is the development of "transpharmers". The gene of a pharmaceutical protein is cloned into livestock animals, and the resulting transgenic animals can be raised for milking the protein.

(2) To manipulate biological pathways. One of the common objectives in gene cloning is to improve crop plants and farm animals. This often involves alteration of biological pathways either by (A) blocking the production of an enzyme, or (B) implementing the production of an exogenous (foreign) enzyme through the manipulation of genes. Many applications of gene cloning in agriculture and food belong to the first category. A well-known example is the inhibition of the breakdown of structural polymers in tomato plant cell wall, by blocking the expression of the gene for the enzyme involved in the breakdown process (using antisense technique). The engineered tomatoes, with decreased softening, can be left to ripe on the vine, allowing full development of color and flavor. Another example is the control of ripening by blocking the expression of the enzyme that catalyzes the key step in the formation of the ripening hormone, ethylene.

On the other hand, new functions can be introduced into plants and animals by introducing a foreign gene for the production of new proteins that are previously not present in the system. The development of pest-resistant plants has been achieved by cloning a bacterial endotoxin. Other examples include salt-tolerant and disease-resistant crop plants. Similar strategies can be applied to raise farm animals, with build-in resistance to particular diseases. Animals cloned with growth hormone genes result in the enhancement of growth rate, increased efficiency of energy conversion, and increased protein to fat ratio to produce lean meat. All these translate into lower cost of raising farm animals, and a lower price for high quality meat.

A number of human genetic diseases, such as severe-combined immunodeficiency (SCID), are caused by the lack of a functional protein or enzyme, due to a single defective gene. In these cases, the defect can be corrected by the introduction of a healthy (normal, therapeutic) gene. The augmentation enables the patient to produce the key protein required for the normal functioning of the biological pathway. "Naked" DNA such as plasmids containing the gene encoding specific antigens can be used as therapeutic vaccines to stimulate immune responses for protection against infectious diseases.

(3) To change protein structure and function by manipulating its gene. One can modify the physical and chemical properties of a protein by altering its structure through gene manipulation. Using the tools in genetic engineering, it is possible to probe into the fine details of how proteins function, by investigating the effects of modifying specific sites in the molecule. This technique has generated vast information on our current knowledge on the mechanism of important proteins and enzyme functions.

In agriculture it is possible to incorporate genetically engineered traits into food crops for the improvement of food quality. The development in this area will enable the production of food ingredients with desirable characteristics directly from food crops, thus eliminating steps currently used in food processing. For example, canola plants may be modified to produce oils with specifications suitable for manufacture of margarine, eliminating the current process of blending and hydrogenation. Also explored are oils as suitable replacements of cocoa butter, the expensive ingredient used in chocolate and candy manufacturing. The nutritional value of food crops can be improved by changing the amino acid contents of seeds. Milk proteins, such as caseins, can be engineered to improve the manufacturing properties in cheese making and other dairy products. Genetically engineered soybean and canola plants produce oils with increased stability and suitable for high temperature frying and low in saturated fat. High solid, low-moisture potatoes that absorb less fat during deep frying, will decrease the amount of fat used for chips and fries. Coffee can be made with better flavor and low caffeine content. The list of applications is endless, the impact of which is covered in Part III and Part IV of this book.

Review

1. Define: (A) a gene, (B) transformation, (C) a clone, (D) expression.
2. What is a vector used for?
3. List some applications of gene cloning.
4. Describe the differences in structural features between prokaryotic and eukaryotic cells.
5. Match by circling the correct answer in the right column.

homozygous dominant	*RR*, *Rr*, *rr*
homozygous recessive	*RR*, *Rr*, *rr*
heterozygous	*RR*, *Rr*, *rr*

6. Tongue rolling is an autosomal recessive trait. What are the genotypes and phenotypes of the children from a heterozygous female married to a homozygous dominant male?
7. Hemophilia is a sex-linked trait. Describe the genotypes and phenotypes of the sons and daughters from a marriage between a normal male and a carrier female.
8. Identify the differences between mitosis and meiosis.

	Mitosis	Meiosis
(A) Number of daughter cells		
(B) Haploid or diploid		
(C) One or two divisions		
(D) Germ cells or somatic cells		

9. Why is it that a dominant allele corresponds to a functional gene? Why is it recessive if a gene is nonfunctional?

CHAPTER 2

STRUCTURES OF NUCLEIC ACIDS

What is the chemical structure of a deoxyribonucleic acid (DNA) molecule? DNA is a polymer of deoxyribonucleotides. All nucleic acids consist of nucleotides as building units. A nucleotide has three components: sugar, base, and a phosphate group. (The combination of a sugar and a base is a nucleoside.) In the case of DNA, the nucleotide is known as deoxyribonucleotide, because the sugar in this case is deoxyribose. The base is either a purine (adenosine or guanine) or a pyrimidine (thymine or cytosine) (Fig. 2.1). Another type of nucleic acid is ribonucleic acid (RNA), a polymer of ribonucleotides also consisting of three components - a sugar, a base and a phosphate, except that the sugar in this case is a ribose, and that the base thymine is replaced by uracil.

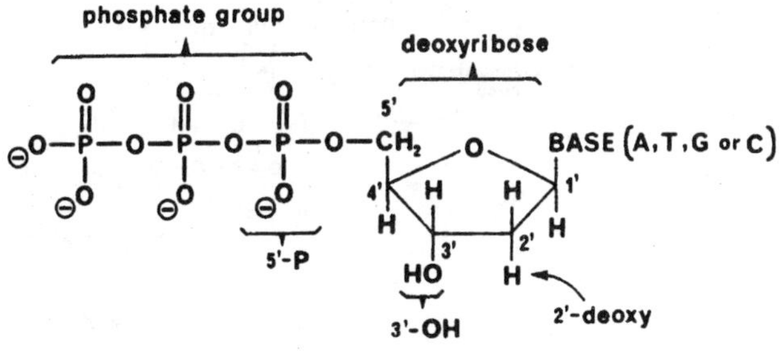

Fig. 2.1. Chemical structure of deoxyribonucleotide

2.1 3'-OH and 5'-P Ends

In DNA, the hydroxyl (OH) group is attached to the carbon at the 3' position. One of the three phosphates (P) in the phosphate group is directly attached to the carbon at the 5' position (Fig. 2.1). The OH group and the P group in a nucleotide are called the 3'-OH (3 prime hydroxyl) and 5'-P. A nucleotide is more appropriately described as 2'-deoxynucleoside 5'-triphosphate to indicate that the OH at the 2' position is deoxygenated and the phosphate group is attached to the 5' position.

A DNA molecule is formed by linking the 5'-P of one nucleotide to the 3'-OH of the neighboring nucleotide (Fig. 2.2). A DNA molecule is therefore a polynucleotide with nucleotides linked by 3'-5' phosphodiester bonds. The 5'-P end contains 3 phosphates but in the 3'-5' phosphodiester bonds, two of the phosphates have been cleaved during bond formation. An important consequence to a phosphodiester linkage is that DNA molecules are directional- one end of the chain with a free phosphate group, and the other end with a free OH group. It is particularly important in cloning to specify the two ends of a DNA molecule: 5'-P end (or simply 5' end) and 3'-OH end (or 3' end).

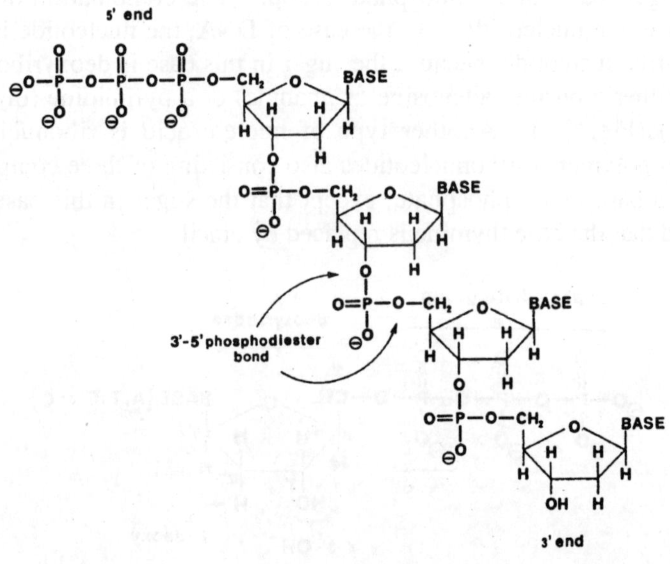

Fig. 2.2. Polynucleotide showing a 3'-5 phosphodiester bond.

2.2 Purine and Pyrimidine Bases

The deoxyriboses and phosphate groups forming the backbone of a DNA molecule are unchanged throughout the polynucleotide chain. However, the bases in the nucleotides vary because there are 4 bases - adenine, thymine, guanine and cytosine, abbreviated as A, T, G and C, respectively (Fig. 2.3). A and G are purines (with double-ring structures); T and C are pyrimidines (with single-ring structures). Consequently, there are four different nucleotides.

Base	DNA Nucleotide (deoxynucleoside triphosphates)	dNTP
Adenine (A)	2'-deoxyadenosine 5'-triphosphate	dATP
Thymine (T)	2'-deoxythymidine 5'-triphosphate	dTTP
Guanine (G)	2'-deoxyguanosine 5'-triphosphate	dGTP
Cytosine (C)	2'-deoxycytidine 5'-triphosphate	dCTP

A DNA molecule with n number of nucleotides would have 4^n possible different arrangement of the 4 nucleotides. For example, a 100 nucleotide long DNA has 4^{100} different possible arrangements. The particular arrangement of the nucleotides (as determined by the bases) of a DNA molecule is known as the nucleotide (or DNA) sequence.

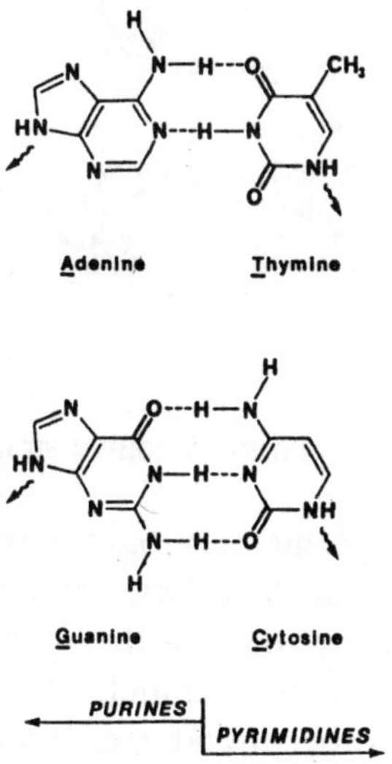

Fig. 2.3. Chemical structures of purine and pyrimidine bases.

2.3 Complementary Base Pairing

The unique structures of the four bases result in base pairing between A and T, and between G and C, by the formation of hydrogen bonds (electrostatic attraction between hydrogen atom and two electronegative atoms, such as nitrogen and oxygen) (Fig. 2.3). It is important to note that there are 3 hydrogen bonds in a GC pair whereas only 2 hydrogen bonds are formed in an AT pair. Therefore, AT pairs are less tightly bound (hence, less stable) than GC pairs.

Base pairing provides a major force for two polynucleotides to interact. A DNA molecule in its native (natural) state exists as a double-stranded molecule, with the nucleotides of one strand base pairing with the nucleotides of the other strand. The two strands in a DNA molecule are therefore complementary to one another. If the bases in one strand are known, the alignment of the bases in the complementary strand can be deduced.

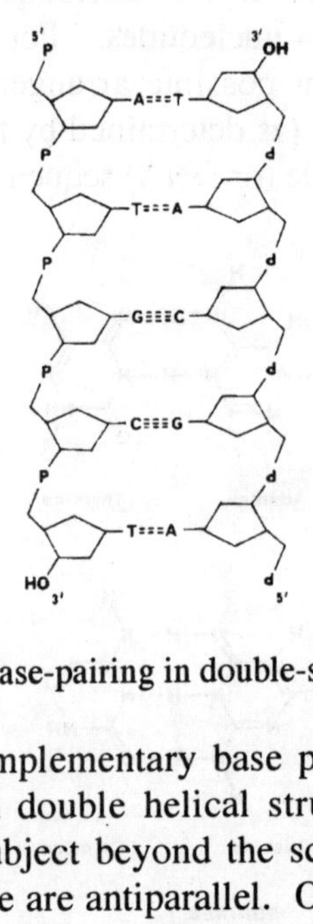

Fig. 2.4. Base-pairing in double-stranded DNA.

In addition to complementary base pairing, the two strands of a DNA molecule assume a double helical structure because of energetic factors of the bonds, a subject beyond the scope of this book. The two strands in a DNA molecule are antiparallel. One strand goes from 5' to 3' in one direction, while the other strand goes in the opposite direction (Fig. 2.4).

2.4 Writing a DNA Molecule

Taking all the information described thus far, a DNA molecule can be represented by a simple scheme. Since the deoxyribose and phosphate backbones are the same for every nucleotide, a DNA molecule can simply be represented by the bases, with indication of the 5' end of the DNA strand. The four bases A, T, G, and C are used synonymously with their respective nucleotides, with the understanding that it is a convenient way to simplify a complicated structure. For example, a DNA sequence is represented: 5'-ATGTCGGTTGA. Also note that a DNA sequence is always read in a 5' to 3' direction. In writing a DNA sequence, always starts with the 5' end. Conventionally, only the sequence of one strand is presented because the complementary strand can readily be deduced. The question then is: Which strand of a DNA molecule do we choose to present? The answer to this is related to the process of transcription and translation, and will be described in Section 4.5.

2.5 Describing DNA Sizes

The size of a DNA molecule is measured by the number of nucleotides (or simply the number of bases). The common unit for double-stranded DNA (dsDNA) is the base pair (bp). A thousand bp is a kilobase (kb). Likewise, a million bp is known as megabase pair (Mb). One kb of dsDNA has a molecular weight of 6.6 x 10^5 daltons (330 gram per mole).

2.6 Denaturation and Renaturation

The two strands of a DNA molecule are held by hydrogen bonds that can be broken down by heating or increasing pH of the DNA solution.

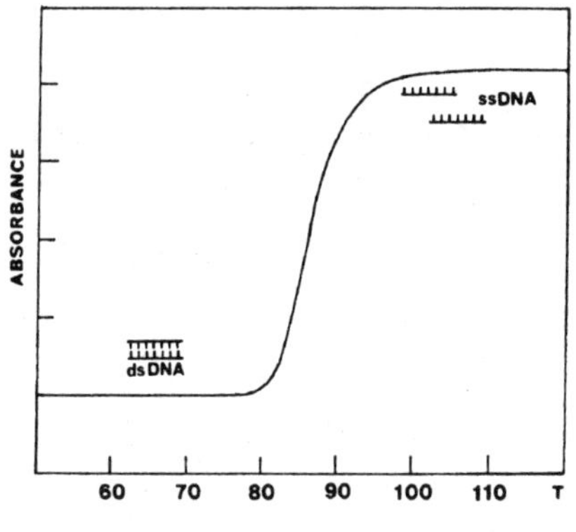

Fig. 2.5. Denaturation and renaturation curve.

In a process known as denaturation, the two strands separate into single-stranded DNA (ssDNA) at a sharp melting temperature ~90°C. Upon cooling of the DNA solution, the two strands reassociate into a dsDNA molecule, a process known as renaturation (Fig. 2.5). The process of thermal denaturation and renaturation is utilized in cloning for creating ssDNA strands, for the annealing of DNA primers in DNA sequencing, and in the polymerase chain reaction (see Sections 8.7 and 8.8).

2.7 Ribonucleic Acid

A second type of nucleic acid is ribonucleic acid (RNA). Like DNA, RNA is also a polynucleotide, but with the following differences (Fig. 2.6): (1) In RNA, the sugar is ribose, not deoxyribose (The nucleotide in RNA is therefore known as ribonucleotide.); (2) The bases in RNA are A, U (uracil), G, C, instead of A, T, G, C in DNA; (3) The OH group at he 2' position is not deoxygenated; (4) RNA is single-stranded. However, it can form base pairs with a DNA strand. For example:

5' - ATGCATG ---- 3'   ssDNA
3' - UACGUAC ---- 5'   RNA

Base	RNA nucleotides(Nucleoside triphospates)	NTP
Adenine (A)	Adenosine 5'-triphosphate	ATP
Uracil (U)	Uridine 5'-triphosphate	UTP
Guanine (G)	Guanosine 5'-triphosphate	GTP
Cytosine (C)	Cytidine 5'-triphosphate	CTP

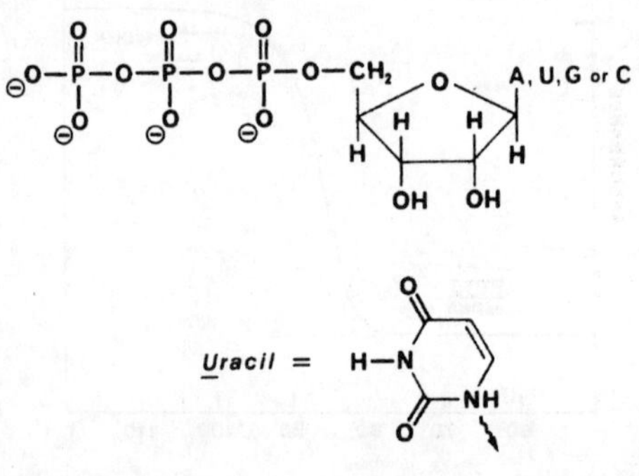

Fig. 2.6. Chemical structure of ribonucleotide.

Review

1. A DNA molecule is formed by linking the _______ of one nucleotide to the _______ of the neighboring nucleotide. The bond formed by linking two nucleotides is a _______________ bond.
2. Deoxyribonucleic acid (DNA) is double stranded. The two strand are _________________ to one another, with A (adenine) pairing with _______ and G (guanine) pairing with _________. DNA strands are directional, with _______ and ______ ends. The two strands are _________ to one another.
3. List the differences in the components between deoxyribonucleotide and ribonucleotide.

Nucleotide	Sugar	Base	Phosphate
Deoxyribonucleotide			
Ribonucleotide			

4. Given the following DNA strand: 5'-TCTAATGGAGCT, write down the complementary strand, ____________________. Indicate the directions by properly labeling the 5' end.
5. What are the conventional rules for writing a DNA sequence?
6. What is the size of the following DNA fragment?

```
5'-AATGGCTAGT GGCAAATGCT AGGCTGCAAG
   CCTTTCCAAT GGTGTGTCAA ACAAAAAACG
   TGCCCGTCAG CAAGTTGTG
```

7. Suppose the DNA fragment in problem 7 is RNA. What will be the sequence like?

CHAPTER 3

STRUCTURES OF PROTEINS

The gene products from transcription and translation are proteins. The structure, and hence the functional property of a particular protein are specified by the information in its gene. Some understanding of the molecular structure of proteins is needed to make sense of the genetic processes.

3.1 Amino Acids

Proteins are polymers of amino acids. There are 20 primary amino acids with a common structure consisting of an amino group (NH_2), a carboxyl group (COOH), and a variable side chain group (R), all attached to a carbon atom (Fig. 3.1, 3.2).

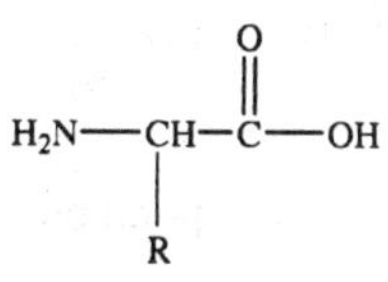

Fig. 3.1. Chemical structure of amino acid.

Each amino acid has a different R group of various chemical structures and physical properties. For example, the amino acid, glycine, has the smallest R group, which is a hydrogen atom. Some amino acids, such as aspartic acid and lysine, have hydrophilic (water-loving) side chains. Some, such as phenylalanine, have hydrophobic characters. Some have side chains that can form charged groups. Amino acid side chains thus can interact in many ways. Important side chain interactions are ionic

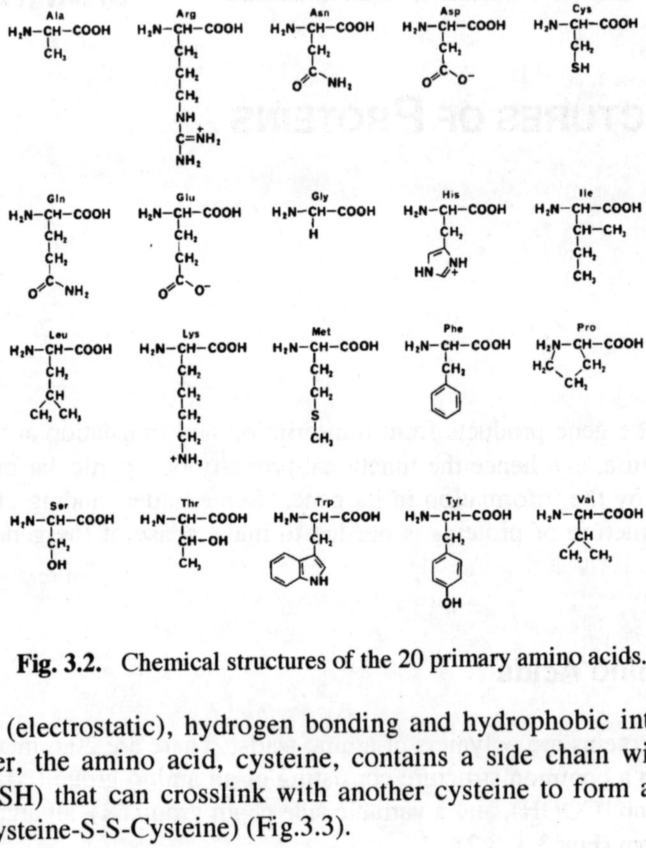

Fig. 3.2. Chemical structures of the 20 primary amino acids.

bonding (electrostatic), hydrogen bonding and hydrophobic interactions. Moreover, the amino acid, cysteine, contains a side chain with a thiol group (-SH) that can crosslink with another cysteine to form a disulfide bond (Cysteine-S-S-Cysteine) (Fig.3.3).

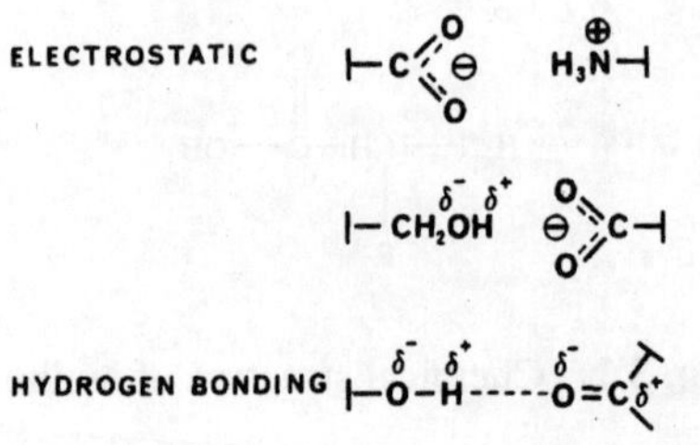

Fig. 3.3. Interactions between amino acid side chains.

Amino acids are frequently represented by 3-letter or 1-letter symbols (Fig. 3.4). For example, alanine is Ala or A; arginine is Arg or R; lysine is Lys or K. One-letter symbols are more often used when presenting the amino acid sequence together with the nucleotide sequence.

Amino acid	3-Letter symbol	1-Letter symbol
Alanine	Ala	A
Arginine	Arg	R
Asparagine	Asn	N
Aspartic acid	Asp	D
Cysteine	Cys	C
Glutamine	Gln	Q
Glutamic acid	Glu	E
Glycine	Gly	G
Histidine	His	H
Isoleucine	Ile	I
Leucine	Leu	L
Lysine	Lys	K
Methionine	Met	M
Phenylalanine	Phe	F
Proline	Pro	P
Serine	Ser	S
Threonine	Thr	T
Tryptophan	Trp	W
Tyrosine	Tyr	Y
Valine	Vale	V

Fig. 3.4. Letter symbols of primary amino acids.

3.2 The Peptide Bond

Proteins are formed by linking amino acids, with the COOH group of one amino acid reacting with the NH_2 group of the succeeding amino acid (Fig. 3.5).

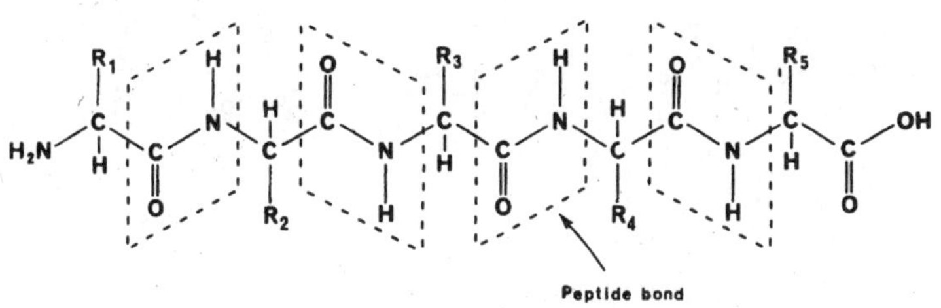

Fig. 3.5. Peptide bond formation.

The linkage formed between two amino acids is a peptide bond, and proteins are polypeptide chains that are directional with N-terminal and C-terminal ends. An amino acid sequence is always written from the N-terminal to the C-terminal end, because proteins are synthesized in this direction. Short chain polypeptides (with fewer than 20 amino acids) are called oligopeptides or simply peptides.

3.3 Structural Organization

The amino acid sequence (the arrangement of amino acids) of a protein is called its primary structure. A protein with n number of amino acids would have 20^n possible different ways of arrangements. From a pool of 20 primary amino acids, a cell can produce thousands of different proteins, each with its specific chemical and biological functions. It is the sequence and the chemical/physical properties of the side chains of the amino acids that define the higher structural shape of a polypeptide chain.

A polypeptide can coil into an α-helix or arrange into pleated sheets, due to the interaction of hydrogen bonds (Fig. 3.6). In an α-helix, the CO groups of each amino acid residue is hydrogen bonded to the NH group of the amino acid residue four units apart. Neighboring amino acids assume a 100° rotation, resulting in 3.6 amino acid residues per turn (360°).

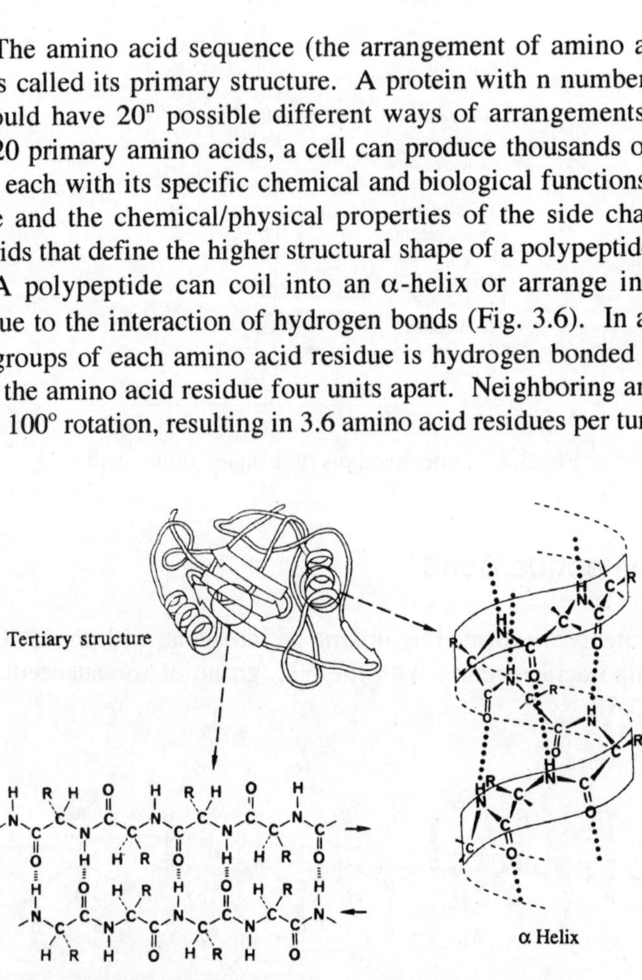

Fig. 3.6. Structural organization of a protein molecule.

In a β-pleated sheet, the polypeptide chains are extended with hydrogen bonds formed between adjacent chains. The alignment of polypeptide chains can be parallel (in the same direction) or antiparallel. These structures are known as the secondary structure of a protein.

Many proteins assume further organization in that the secondary structure folds back to form compact globular structure as a consequence of interactions such as hydrogen bonding, hydrophobic forces, ionic interactions and disulfide crosslinks among the amino acid side chains. Interactions between the side chains and the water molecules in the immediate environment of the protein also play a major contribution in the process. These structural arrangements describe a protein's tertiary structure. Certain proteins consist of more than one polypeptide. In this case, two or more polypeptides assemble to form a large molecule. For example, hemoglobin, an important protein found in red blood cells that reversibly binds oxygen, is a tetramer. Each of the four polypeptides is a subunit of the overall structure; each subunit processes similar secondary and tertiary structures. The assembly of subunits is called quaternary structure. In nature, not all proteins assume a globular shape. For example, collagen, the protein that provides mechanical strength to bones, cartilages, and skins, consists of three polypeptides interweaved to form a triple-helical rod-shape structure.

3.4 Posttranslational Modification

Proteins may undergo a number of modifications after they are synthesized. Proteins are often synthesized with an extra short peptide in the N-terminal end, which will be cleaved at a later stage. This N-terminal peptide sequence may serve to keep the protein in nonfunctional form until it is activated into the mature active form. This provides a precise control in the timing and location for the action of a particular protein in the physiological processes of a cell. Some proteins are synthesized and secreted from the cell. In this case, the short sequence is a signal peptide that functions to guide the protein through various compartments in the cell to the outer surface of the plasma membrane.

Many proteins exist as glycoproteins or lipoproteins. The former has carbohydrates covalently attached to the protein molecule, whereas the latter has lipid molecules attached. The addition of carbohydrate or lipid components to a protein occurs after the translation process. Other modifications include phosphorylation (adding phosphate groups to amino acid side chains), and acetylation (adding acetyl groups).

3.5 Enzymes

Enzymes are a special class of proteins that function to accelerate biochemical reactions in cells. Without enzymes, few reactions in biological systems can occur. The chemistry involved in the mechanism of accelerating a reaction is called catalysis. An enzyme catalyzes a specific chemical reaction without itself being consumed. In an enzymatic process, the starting chemical (called substrate) is converted to a new compound (product) in a rate million times faster than the uncatalyzed reaction, often at low temperature and near neutral pH. The enormous rate enhancement in enzyme catalysis is made possible by the formation of an enzyme-substrate complex. The enzyme binds its substrate at a position optimal for the reaction to proceed. The location in an enzyme molecule where the substrate binds and catalysis occurs is the enzyme's active site. The proximity effect results in lowering the energy required for the reaction to occur.

In any chemical reaction, the direction of equilibrium is described by ΔG, the change in the free energy of the reaction. In a chemical reaction A + B = C + D, if reactants A and B possess more free energy than the product, C and D, than ΔG (which is equal to $G_{products} - G_{reactants}$) becomes negative, and the reaction proceeds to the right. Increasing ΔG will shift the equilibrium increasingly to product formation. Likewise, if ΔG is positive, the reaction will not proceed, as the equilibrium is favored to the left.

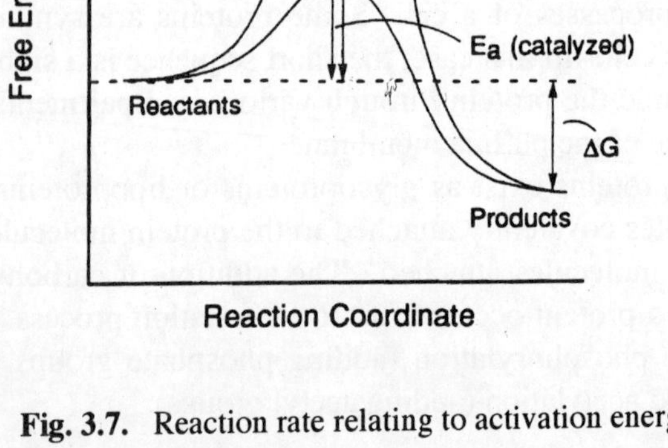

Fig. 3.7. Reaction rate relating to activation energy.

The ΔG of a reaction describes only the equilibrium position of the reaction; it does not describe how *fast* the reaction goes in attaining the equilibrium position. This rate of reaction is related to the activation energy, E_a, a measure of the energy barrier that represents the formation of the transition state between reactants and the products. The height of this energy barrier determines the rate of a reaction at a given temperature (Fig. 3.7). In enzyme-catalyzed reactions, an enzyme lowers the activation energy for the transition state of the substrate by the formation of an enzyme-substrate complex. The transition state is an unstable species in which bonds are constantly forming and breaking. The binding of an enzyme to a substrate occurs with complementarity of conformational shapes that imparts substrate specificity – a unique characteristic of enzyme actions. No molecules, other than its specific substrates or analogs, can form complex with an individual enzyme. When the substrate binds with an individual enzyme, it often induces a change in the conformation of the enzyme where the active site is properly poised for catalysis (Fig. 3.8).

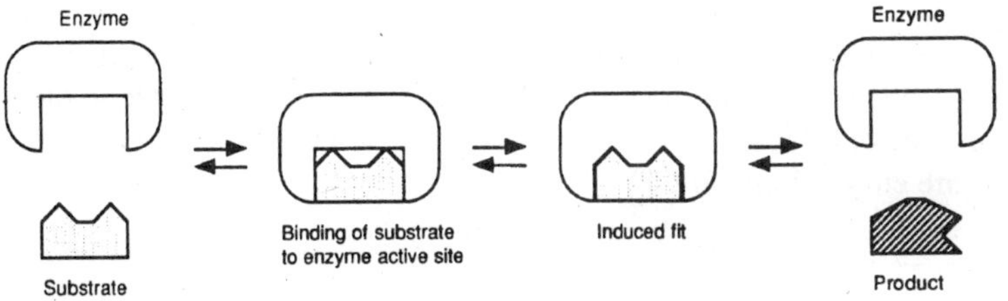

Fig. 3.8. Illustration of enzyme-substrate interaction in catalysis.

Enzymes are divided into 6 classes according to the system recommended by the Commission on Enzymes of the International Union of Biochemistry. The enzymes used in cloning can be classified into one of the six groups, and several of them are hydrolases.

1. Oxidoreductases - oxidize or reduce substrates.
2. Transferases - remove groups from one substrate and transfer them to an acceptor molecule.
3. Hydrolases - catalyze the breakage of covalent bonds with the concurrent addition of water.
4. Lyases - remove groups from substrates to leave a double bond, or adding groups to double bonds.
5. Isomerases - catalyze isomerization of substrates.
6. Ligases - catalyze bond formation accompanied by the breaking of ATP or similar triphosphates.

Review

1. Proteins are polymers formed by linking ______ group of one amino acid to the ______ group of the preceding amino acid, forming a ______ bond.
2. There are ______ amino acids, each represented by a three-letter or one-letter symbols. The number of possible arrangements for a protein is ______. For a peptide consisting of eight amino acids, the possible number of combinations of different arrangements is equal to ______. (What will be the answer if this is a DNA fragment of eight nucleotides?)
3. Define: primary structure, secondary structure, tertiary structure, and quaternary structure, of a protein. What are the major forces involved in the formation of each structural organization?
4. Give examples of pairs of amino acids that form (A) electrostatic interations, (B) hydrogen bonding, (C) hydrophobic interactions. (Refer to Figs. 3.2 and 3.3)
5. What are the conventional rules for writing a protein sequence?
6. Enzymes are proteins with special functions of ______________.
7. ΔG describes the change in the free energy of a reaction. If ΔG is ______, a reaction will not occur. For the formation of products, the value of ΔG is ______.
8. The rate of an enzyme-catalyzed reaction is determined by the activation energy of the reaction. What is activation energy? How is it related to the transition state of a substrate, and the formation of an enzyme-substrate complex?
9. How are enzymes classified?

CHAPTER 4

THE GENETIC PROCESS

Two processes are central to genetic continuity from one generation to the next: (1) Genetic information is conveyed from DNA to RNA to proteins (transcription and translation); (2) Genetic information is transferred from DNA to DNA (replication).

4.1 From Genes to Proteins

The genetic information carried by DNA is expressed in the form of proteins by a two-stage process. The first is transcription (DNA —>mRNA) in which the information (nucleotide sequence) in the DNA is transcribed into messenger RNA (mRNA). The second is translation (mRNA —> protein) in which the mRNA sequence is decoded (translated) into an amino acid sequence. The following sections describe the general scheme of the process.

4.2 Transcription

In the synthesis of mRNA, only one of the two DNA strands is transcribed. The DNA strand that is used in transcription is called the template strand (Fig. 4.1). Transcription requires the action of an enzyme called RNA polymerase which recognizes and binds to a segment of DNA preceding the 5' end of the gene.

In an initial step, the dsDNA unwinds at the binding of the RNA polymerase. The mRNA is synthesized in a 5' to 3' direction with the bases (ribonucleotides) forming complementary pairs (i.e., forming AU and GC pairs) with those of the template DNA strand. The ribonucleotides in the developing mRNA are linked in a polymerization reaction whereby the 3'-OH of one nucleotide reacts with the 5'-P of the succeeding nucleotide. The RNA polymerase moves along the DNA, unwinding, base-pairing, and polymerizing the growing mRNA until the termination site is reached. The mRNA, as it is formed, separates from the DNA template strand, allowing that portion of the DNA to rewind.

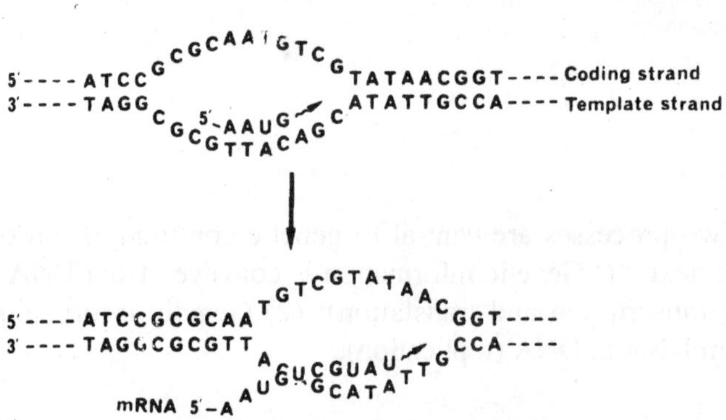

Fig. 4.1. Template DNA strand transcribed into mRNA.

It is important to note that the synthesized mRNA is complementary to the DNA template strand. The DNA template strand is also known as anticoding, noncoding, or antisense DNA strand. The DNA strand complementary to the template strand bears the same sequence as the mRNA, and is called the coding, or sense DNA strand. The term "transcript" is sometimes used to describe an RNA copy of a gene.

4.3 Translation

The translation of mRNA requires two additional classes of RNA: (1) Ribosomal RNA (rRNA) which forms a major component of the ribosome where translation occurs; (2) Transfer RNA (tRNA) which "reads" the mRNA and converts the information in the nucleotides into an amino acid sequence. Transfer RNA assumes a cloverleaf structure, with the 3' end attached to an amino acid, and a loop region consisting of a 3-nucleotide anticodon. There are 20 amino acids, each carried by one or more tRNAs with specific anticodons.

Translation is initiated by the attachment of mRNA to a ribosome. The nucleotide sequence of mRNA is translated into protein in a 5' to 3' direction. The ribosome moves along the polynucleotide chain as translation proceeds (Fig. 4.2). At the ribosome, mRNA is read by tRNAs every 3 successive nucleotides. Each group of three successive nucleotides in the mRNA constitutes a codon that base-pairs with the anticodon of an individual tRNA which carries a specific amino acid. The amino acid carried by the paired tRNA links to the neighboring amino acid in the developing polypeptide as the ribosome moves to the next codon in the mRNA. The NH_2 group of one amino acid forms a peptide bond with the COOH group of the preceding amino acid. The synthesis of proteins proceeds from the N- to the C-terminus, as the mRNA is read by tRNAs from the 5' to the 3' end.

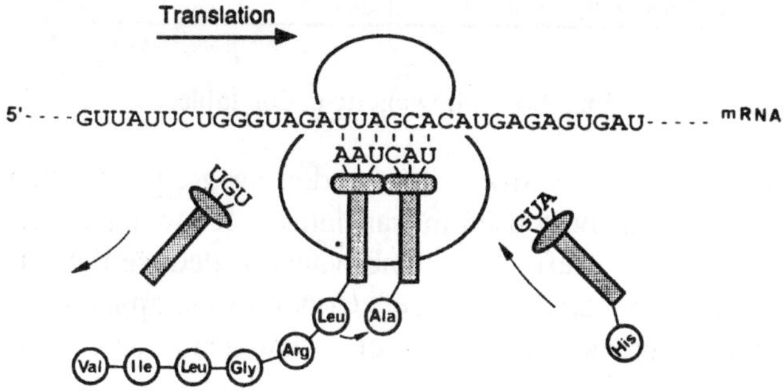

Fig. 4.2. Translation of mRNA involves ribosomal RNA and transfer RNA.

4.4 The Genetic Code

There are 64 possible codons (3 nucleotides in each codon with 4 possible base, total of $4^3 = 64$ possible codons). Of the 64 codons used to code for the 20 amino acids, 1 codon (AUG) is used as a start signal for translation, and 3 codons (UAA, UAG, or UGA) are termination signals. The AUG codon also codes for the amino acid methionine. Since there are 61 codons for 20 amino acids, there is more than one codon coding for one amino acid, as can be seen clearly from the codon table (Fig. 4.3)

For example, Phe is coded by 2 codons, either UUU or UUC in the mRNA sequence. This means that the tRNA carrying Phe at its 3' end has its anticodon either AAA or AAG. Notice that the codon table refers to the codons as present in mRNA, not the anticodons in tRNA. In other words, using the codon table, one can translate the mRNA directly into amino acids.

5' end	Second Position: U	C	A	G	3' end
U	UUU Phe	UCU Ser	UAU Tyr	UGU Cys	U
	UUC Phe	UCC Ser	UAC Tyr	UGC Cys	C
	UUA Leu	UCA Ser	UAA Stop	UGA Stop	A
	UUG Leu	UCG Ser	UAG Stop	UGG Trp	G
C	CUU Leu	CCU Pro	CAU His	CGU Arg	U
	CUC Leu	CCC Pro	CAC His	CGC Arg	C
	CUA Leu	CCA Pro	CAA Gln	CGA Arg	A
	CUG Leu	CCG Pro	CAG Gln	CGG Arg	G
A	AUU Ile	ACU Thr	AAU Asn	AGU Ser	U
	AUC Ile	ACC Thr	AAC Asn	AGC Ser	C
	AUA Ile	ACA Thr	AAA Lys	AGA Arg	A
	AUG Met	ACG Thr	AAG Lys	AGG Arg	G
G	GUU Val	GCU Ala	GAU Asp	GGU Gly	U
	GUC Val	GCC Ala	GAC Asp	GGC Gly	C
	GUA Val	GCA Ala	GAA Glu	GGA Gly	A
	GUG Val	GCG Ala	GAG Glu	GGG Gly	G

Fig. 4.3. The genetic codon table.

Many organisms show unique codon usage, in that there is a preferred set of codons heavily used in translation. Knowing codon usage of a particular organism is useful when one wants to deduce the nucleotide sequence from an amino acid sequence. Coden usage apparently affects the protein synthesis and secretion, a factor that needs to consider in gene cloning.

4.5 Why Present a Sequence Using the Coding Strand?

By convention, a DNA sequence is described by the coding strand. It is common practice to write down the coding strand in presenting a DNA sequence, although the template strand is used for base pairing in transcription in the biological process. The reason is that because in cloning, we are interested in the sequence of the mRNA, which is a copy of the coding strand and not that of the template strand.

```
5′ ---- TGGTTTACCTCT ----
3′ ---- ACCAAATGGAGA ----
```

Suppose the top strand is the coding strand, then for the nucleotide sequence of the mRNA, simply copy down the same DNA sequence except replacing the T by U as follow.

```
5′ ---- UGGUUUACCUCU ----
```

For translation, what we need to know is the mRNA nucleotide sequence. The amino acids can be read directly from the codon table. It is not necessary to deduce the anticodons of tRNAs although this step does occur in the biological system. Thus, translation of the above mRNA gives the following amino acid sequence.

```
5′ ---- UGG UUU ACC UCU ----mRNA
N  ---- Trp Phe Thr Ser ----Amino acids
```

In fact, one can write down the amino acid sequence directly from reading the coding strand, taking care that the T bases are replaced by U. The conversion can be conveniently done by the use of computer software readily available in the public domain. It is also relatively easy to reverse the process, deriving DNA sequence from a known amino acid sequence. In this process, because each amino acid is coded by more than one codon, codon usage for a particular organism must then be taken into consideration.

4.6 The Reading Frame

The process of transcription/translation implies that it requires a number of precise control elements. First, there must be a way to distinguish which strand of the DNA molecule is transcribed. The mRNA nucleotide sequence is different depending on which one of the two DNA strands is used as the template. Consequently, the translation of these two mRNAs will give two proteins with completely different amino acid sequences.

```
5′ ---- TGGTTTACCTCT ----
3′ ---- ACCAAATGGAGA ----
```

(1) Top strand = coding strand,

```
5′ ---- UGG UUU ACC UCU ----mRNA
N  ---- Trp Phe Thr Ser ----Amino acids
```

(2) Lower strand = coding strand,

```
5′ ---- AGA GGU AAA CCA ----mRNA
N  ---- Arg Gly Lys Pro ----Amino acids
```

Second, there must be precise controls of the start site and termination site for both transcription and translation. The genetic code is read in groups of every 3 nucleotides. A shift in the reading frame will result in a different protein. Transcription can also affect indirectly the reading frame. Different transcription start sites give mRNAs with different 5' end, which will cause a shifting of the reading frame in translation.

```
5' ---- TGGTTTACCTCT ---- Coding strand
         a b          ← Transcription start site
```

Transcription with start site at a,

```
5' UGG UUU ACC UCU ----mRNA
N  Trp Phe Thr Ser ----Amino acids
```

Transcription with start site at b,

```
5' GUU UAC CUC U ----mRNA
N  Val Tyr Leu ----   Amino acids
```

For translation, starting with the same mRNA, but different translation start sites, also results in a different reading frame.

```
5' ----- UGGUUUACCUCU ----- mRNA
          a b         ← Translation start site
```

Reading frame with start site at a,

```
5' ---- UGG UUU ACC UCU ----mRNA
N  ---- Trp Phe Thr Ser ----Amino acids
```

Reading frame with start site at b,

```
5' ---- UG GUU UAC CUC U ----mRNA
N ----     Gly Leu Pro ----   Amino acids
```

In gene cloning, it is importance, if expression of the gene is desired, to insure that the gene is properly inserted into a vector so that it is placed in the correct reading frame. In frame or out-of-frame is one of the factors in determining success or failure of gene expression. Construction with a correct frame often requires knowing the sequence, in particular the 5' end portion of the gene, and paying careful attention to making the DNA insertion.

The term "open reading frame" is often used in gene cloning. A DNA sequence is read (often using computer software) to eliminate the frames that are interrupted by stop codons. The reading frame that yields complete translation of the entire sequence without interruption is called the "open reading frame" of that sequence. In the following example, only (B) gives an open reading frame.

```
5′ --- TTCTCAGTTAATTAATGTAGT ---
```

(A) Reading starts at the first T.

`N --- Phe Ser Val Asn` Stop

(B) Reading starts at the second T.

`N --- Ser Gln Leu Ile Asn Val`

(C) Reading starts at C.

`N --- Leu Ser` Stop

4.7 DNA Replication

The discussion thus far describes the conversion of DNA information for the synthesis of proteins. The discussion is incomplete without consideration of another important process, DNA replication. Replication is the process whereby a DNA molecule duplicates to yield identical DNA molecules. The duplication of genetic materials is an essential part of cell division, so that the daughter cells produced in cell division carry the same genetic information as the parent cell.

In replication, the dsDNA unwinds by the action of an enzyme called helicase. The resulting Y-shape structure is called a replication folk (Fig. 4.4). The basic features of replication are the same for both strands. Nucleotides are added complementary to either of the two strands. Phosphodiester bonds are formed by the action of the enzyme, DNA polymerase III. In order for the enzyme to work, a short RNA primer complementary to the parent DNA strand is needed. This RNA primer is formed by an enzyme, RNA primase, which is part of a larger enzyme complex known as primosome.

The two strands at the replication fork are different in that one strand has an exposed 3' end, and the other an exposed 5' end. For the strand with exposed 3' end, the replication proceeds continuously. The new daughter strand is synthesized in a 5' to 3' direction. This daughter strand is known as the leading strand.

For the strand with the exposed 5' end, replication proceeds in a discontinuous fashion, in short segments, in a 5' to 3' direction. The short segments, known as Okazaki fragments, are later joined together to form a complete daughter strand. This daughter strand is called the lagging strand. The joining of Okazaki fragments requires the action of DNA polymerase I and DNA ligase. The action of DNA polymerase I is to replace the RNA primer with DNA extension from the upstream Okazaki fragment. When the entire RNA primer is replaced, the gap separating the two Okazaki fragments is joined by the formation of a phosphodiester bond by the action of DNA ligase.

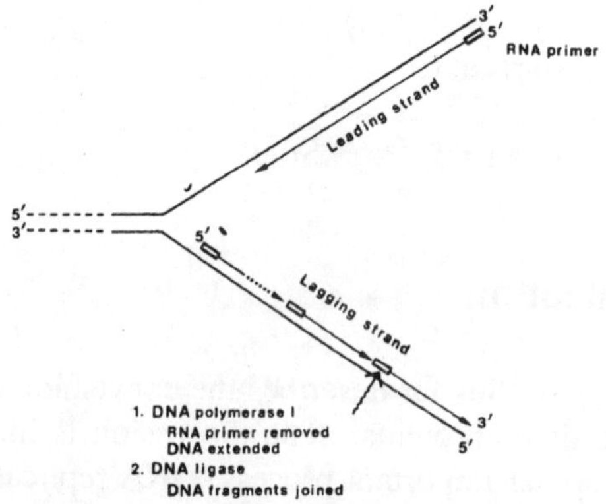

Fig. 4.4. Replication fork showing synthesis of leading and lagging strands.

4.8 The Replicon and Replication Origin

The unit of DNA replication is called the replicon. In the case of bacteria, the entire genome constitutes a single replicon. Replication starts from an initiation point (called origin of replication) with the formation of a replication fork and proceeds until the entire genome is completely duplicated. Replication may be unidirectional or bidirectional, depending on the movement of the replication fork at the origin. The frequency of replication is dependent on the complex control of regulatory proteins with the origin. Bacterial plasmids, bacteriophages, or virus DNA contain replicons which may be described by a similar scheme of replication. Eukaryotic chromosomes, however, consists of many replicons which are generally smaller and replicate at a lower rate than bacterial replicons.

The complete DNA sequence of a replication origin can be isolated, cloned into a DNA molecule lacking an origin, and the resulting recombinant DNA acquires the ability to replicate. Replication origins isolated from bacteria have A-T rich sequences. One of the key features of a cloning vector is an appropriate replication origin to ensure proper replication of the inserted gene (See Section 9.1).

4.9. Relating Replication to Gene Cloning

In cloning, it is often necessary to obtain a DNA segment in quantities sufficient for handling and manipulation. The biological process of replication has been the primary choice of producing DNA in large quantities. The desirable DNA is inserted into a vector, which is then introduced into bacterial cells, such as *E. coli*. Transformed cells are cultured, harvested, and lyzed. The DNA released from the cells can be isolated and purified.

The action of DNA polymerase also forms the basis of the polymerase chain reaction (PCR) that enables the amplification of a chosen region of a DNA molecule, as long as the flanking regions are known (see Section 8.9). The enzyme responsible for DNA replication in *E. coli* is DNA polymerase III, which requires a RNA primer for action. The DNA polymerase used in cloning work is *E. coli* DNA polymerase I which requires a DNA primer and possesses 5'->3' exonuclease activity (see Section 7.3).

Review

1. Describe a complete set of rules for writing a DNA sequence.
2. Given the DNA strand: 5'-TCTAATGGAGGT, the complementary strand is ______________________. If the complementary strand is the template, the mRNA reads _________________________ . If the above mRNA is used for translation, the amino acid sequence is ____________.
3. Repeat problem 2, using the given DNA strand as the template.
4. Repeat problem 2, but start transcription at the second base of the mRNA.
5. Repeat problem 2, but start translation at the second base of the mRNA.
6. What is an open reading frame? What is the open reading frame for the following sequence?

5'-TCTTGTAATTGACGTCGGAAT

7. Why is it that replication of the strand with the exposed 5' end proceeds in a discontinuous manner?
8. Can you suggest one reason why the sequences of bacterial replication origins are A-T rich?

CHAPTER 5

ORGANIZATION OF GENES

A gene is a discrete segment of DNA existing as an expression unit (see Section 1.1). A DNA molecule may consist of many genes. For example, in bacteriophage λ, all the genetic information is stored in a single DNA molecule with 48.5 kb, consisting of 60 genes. In humans, the genetic information is stored in 46 DNA molecules organized as 23 pairs of chromosomes, amounting to a total of 3.2×10^9 bp, and estimated ~31,000 genes (see Section 1.4).

Consider a general organization of a structural gene in a DNA molecule. The DNA segment preceding the transcription start site is the 5' flanking region. This is also known as the upstream region. The DNA sequence following the transcription termination site is called the 3' flanking region or downstream region. The terms "upstream" and "downstream" are also used to refer the relative positions of two locations in a sequence.

5.1 The Lactose Operon

When and where transcription of the template strand to mRNA occurs is precisely controlled. The start and termination signals for transcription are controlled by a set of regulatory elements located in the upstream and downstream regions of a structural gene. The collection of the control regions and the structural gene(s) is called an operon.

A well- known example is the lactose (*lac*) operon in *E. coli* bacteria. The *lac* operon consists of the following elements - regulatory gene (s), promoter, operator, and structural gene(s) (Fig. 5.1). In prokaryotes, it is not uncommon that a single regulatory mechanism controls more than one structural gene. The *lac* operon is a typical example; it consists of 3 structural genes, *lacZ*, *lacY*, and *lacA*, coding for the three enzymes,

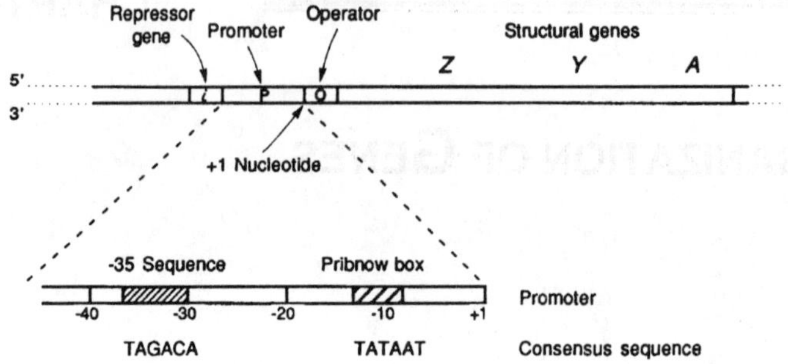

Fig. 5.1. Organization of the lactose operon.

β-galactosidase, permease, and acetylase, respectively, all under the control of one regulatory mechanism. The control region located upstream of these 3 structural genes consists of the repressor gene, promoter and operator. The control regions are sometimes collectively referred to as the "promoter region"; the term "promoter" here is used in a loose sense.

5.2 Control of Transcription

There are two types of transcriptional controls: location and timing. (1) Where are the start site and the termination site for transcription? (2) When does transcription start or stop?

5.2.1 The Transcription Start Site and Termination Site

Transcription starts with RNA polymerase recognizing the promoter sequence as the binding site. RNA polymerase is a holoenzyme consisting of the core enzyme plus sigma factors. The later are proteins that assist in the recognition of the promoter by the enzyme. Specific sigma factors are responsible for directing RNA polymerase to specific promoters. All promoters consist of common short sequences recognized by RNA polymerase. These are called consensus sequences. Two consensus sequences are located in *E. coli* promoters, known as the -35 sequence (5'-TTGACA) and the -10 sequence or Pribnow box (5'-TATAAT) (Fig. 5.1). The binding of RNA polymerase to the promoter determines the start site for transcription as well as which strand to copy. Figure 5.2 shows the *lac* promoter sequence bound by RNA polymerase, and the unwinding of the DNA strands by the enzyme. The transcription start site, conventionally named as the +1 nucleotide, is located at the beginning of the operator sequence in the *lac* operon as indicated in the illustration. (The position of +1 nucleotide varies in different operons. For example, in the *trp* operon, the +1 nucleotide is down stream of the operator.) The sequence upstream of the +1 nucleotide is numbered as the minus (-) nucleotides.

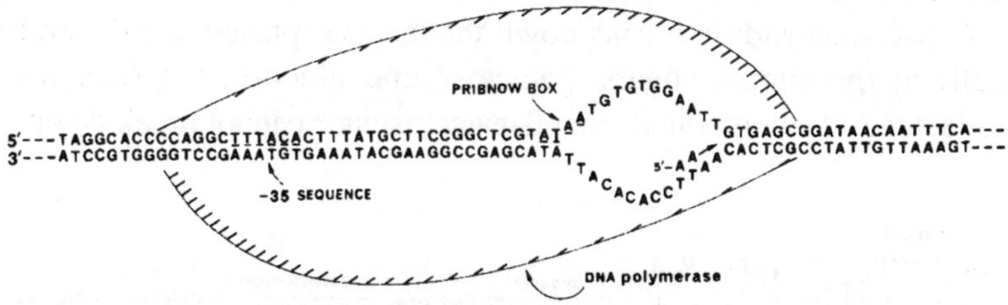

Fig. 5.2. Binding of RNA polymerase to the *lac* promoter.

Transcription termination sites consist of sequence with invert repeats of high GC pairs. Consequently the segment of mRNA transcribed in this region folds into a stem-loop (Fig. 5.3). The RNA polymerase, when it comes past the stem-loop structure, detaches from the DNA strands. In some termination sites, the detachment is assisted by a Rho protein which functions to break DNA-RNA base pairs between the template and the mRNA.

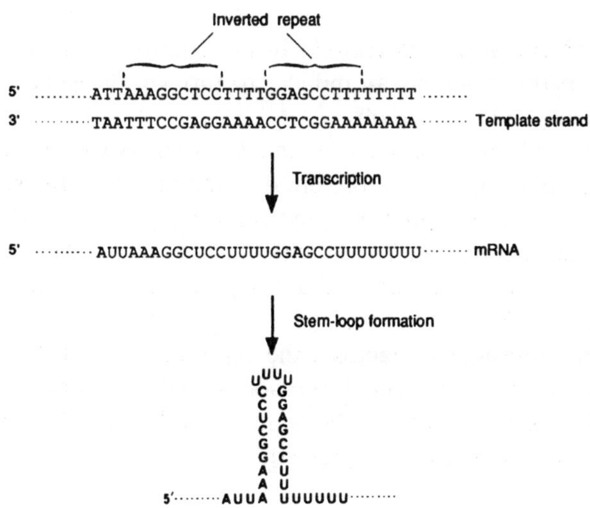

Fig. 5.3. Formation of the stem-loop structure.

5.2.2 Where Does Transcription Start or Stop?

The process of protein synthesis is controlled at the transcriptional step. Proteins are synthesized only when needed. This turn-on and turn-off of a gene are well illustrated in the *lac* operon. The functions of the three enzymes coded by *lacZ*, *lacY*, and *lacA* are: (1) Permease facilitates the active diffusion of lactose from the medium into the bacterial cell (Fig. 5.4); (2) β-Galactosidase breaks down the lactose (present inside the bacterial cell) to the simple sugars, galactose and glucose; (3) Acetylase removes lactose-like compounds that β-galactosidase cannot break down.

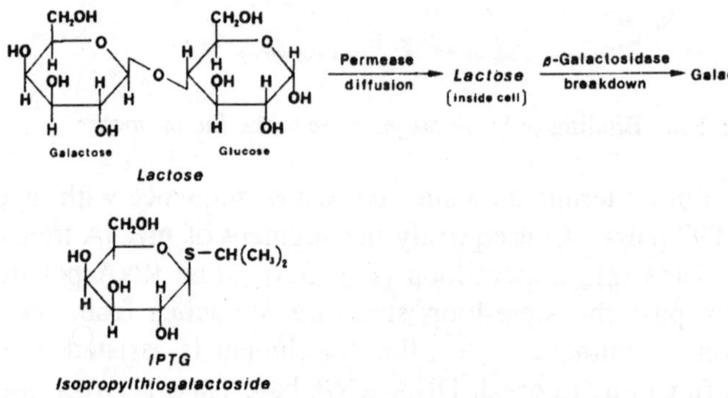

Fig. 5.4. Reaction catalyzed by permease and β-galactosidase.

Lactose as an Inducer. In the control mechanism, lactose in the growth medium acts as an inducer to turn on the genes. Its absence causes the genes to be turned off.

In the absence of lactose, the *lac* repressor, the protein of the regulatory gene in the *lac* operon, binds to the operator (the short DNA sequence between the *lac* promoter and the *lacZ* gene). Binding of the repressor to the operator interferes with the interaction between RNA polymerase and the promoter. Transcription cannot occur under this condition (Fig. 5.5).

In the presence of lactose, the repressor protein binds to lactose and the resulting lactose-bound repressor is unable to bind to the operator. The promoter is accessible to the binding of RNA polymerase, and the *lacZ*, *lacY* and *lacA* genes are expressed.

Glucose as a suppressor. The *lac* operon also has a positive control mechanism. Transcription occurs only when an activator, which is a complex of cAMP-CRP (cyclic AMP receptor protein), binds to the promoter. The concentration of cAMP increases only when glucose is not available in the cell. When glucose (which is the product of hydrolysis of lactose by β-galactosidase) is present, the level of cAMP decreases, and the activator is not functional. In this case, glucose acts as a suppressor to turn off the gene (Fig. 5.5).

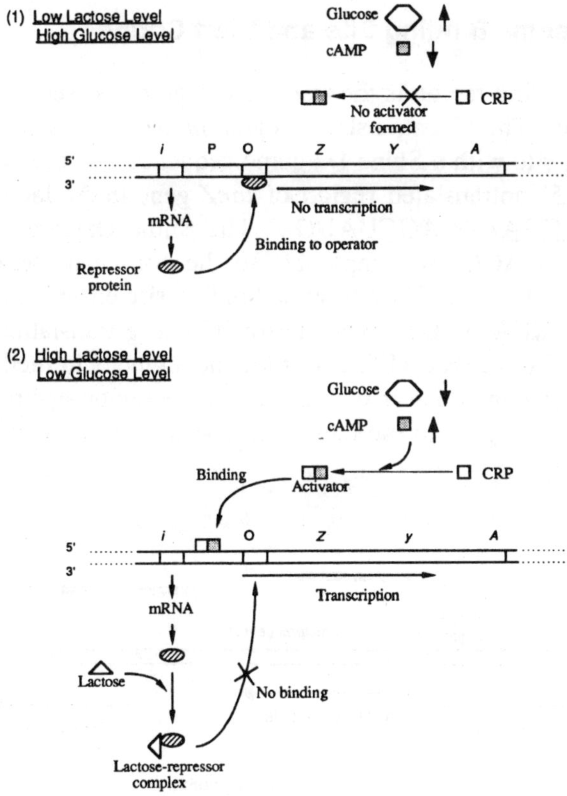

Fig. 5.5. The regulatory process of the *lac* operon.

In gene cloning, the control region (regulatory gene, promoter, operator) is utilized as a genetic switch. In a simple scheme, a controlled expression system can be constructed by putting the genetic switch upstream of any gene (see Section 9.1.1, Fig. 9.2). The natural inducer, lactose, is replaced by isopropylthiogalactoside (IPTG) which is not hydrolyzed by β-galactosidase. Gene expression is initiated by the addition of IPTG to the culture medium. The gene is turned off if the medium is deprived of IPTG.

5.3 Control of Translation

The mRNA copied from a gene is not translated for its entire sequence. The translation process needs to have a start site and a termination site located, respectively, at the 5' end and the 3' end regions of the mRNA.

5.3.1 Ribosome Binding Site and Start Codon

The coding region is the translated region that specifies the amino acid sequence. The 5' untranslated region of the mRNA contains a ribosome binding site with a Shine Dalgarno sequence that can base pair with rRNA. The 5' untranslated region of *lacZ* gene in the lac operon is 5'-UUCACCCAGGAAACAGCUAUG-. The Shine Dalgarno sequence in this instance is AGGAA comparable to the consensus sequence of 5'-AGGAGGU in *E. Coli*. The ribosome binding site ensures the correct positioning of mRNA on the ribosome for initiating translation at the start codon AUG. Notice that AUG codes for the amino acid methionine. The A in the start codon is designated as +1 and the region upstream as minus (-) sequence, analogous to the numerical coordinates used in transcription (Fig. 5.6).

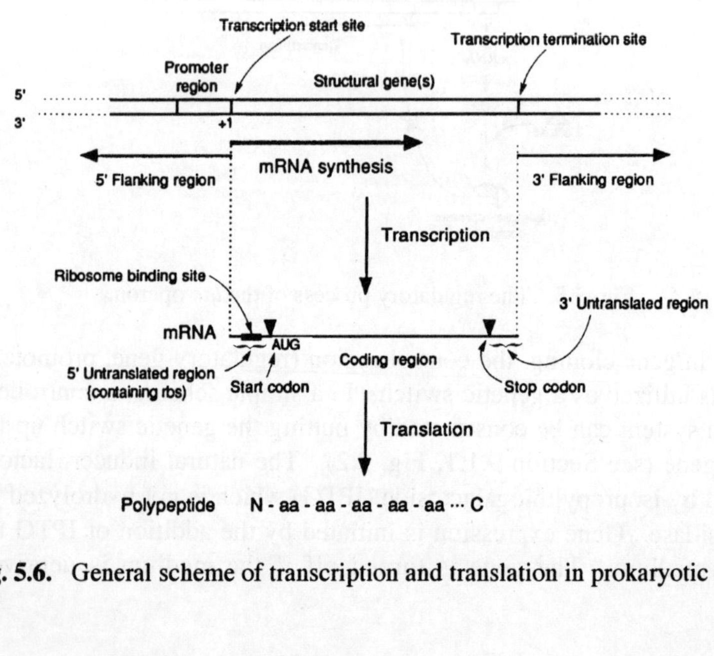

Fig. 5.6. General scheme of transcription and translation in prokaryotic cells.

5.3.2 Translation Termination Site

Translation terminates at a stop codon (AGC, UAG, or UAA). There are no tRNAs with anticodons that can base pair with these three codons. The mRNA therefore detaches from the ribosome. In some cases, arrangement of several stop codons may act as the termination signal.

5.4 The Tryptophan Operon

Another well-studied operon in *E. coli* is the *trp* operon, which is involved in the biosynthesis of the amino acid, tryptophan. The synthesis of tryptophan requires 5 enzymes, encoded by 5 genes (*trpE*, *trpD*, *TrpC*, *TrpB*, and *TrpA*). All 5 genes are controlled under a single regulatory system. The following control elements are involved: (1) a *trp* promoter and an operator, (2) a repressor gene (*trpR*) located distant from the cluster of *trp* genes, (3) a leader (*trpL*) of 162 nucleotides located between the promoter region and the *trpE* gene (the first gene in the *trp* operon) (Fig. 5.7).

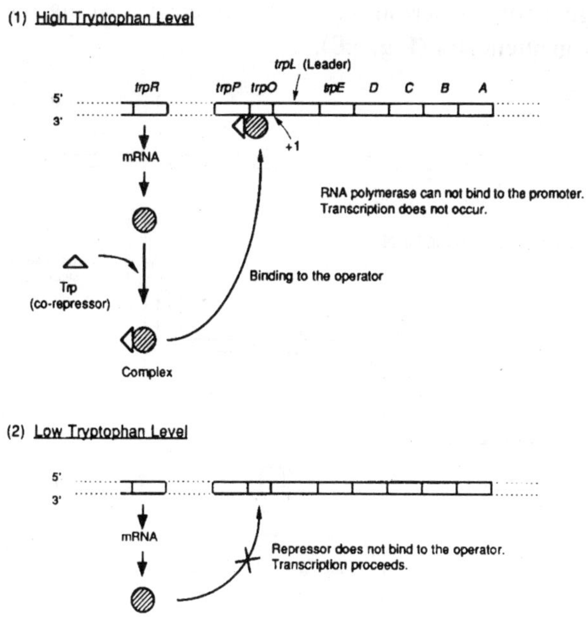

Fig. 5.7. Organization of the tryptophan operon.

5.4.1 Co-repressor

Transcription of the *trp* genes is under the control of the *trp* repressor. In the presence of high concentrations of intracellular tryptophan, the *trp* repressor protein binds to the amino acid to form a repressor-tryptophan complex. Under this condition, the enzyme RNA polymerase is prevented from binding to the promoter and transcription cannot occur. In the absence of tryptophan, the *trp* repressor protein on its own cannot bind to the operator, and transcription can proceed.

In contrast to the *lac* operon in which lactose acts as an inducer, the amino acid tryptophan is the co-repressor in this system. A complementary regulatory activity called attenuation is also involved.

5.4.2 Attenuation

The leader sequence consists of 4 regions (R1, R2, R3, and R4) capable of base-pairing to form a variety of loop structures: (1) R1 pairs with R2, and R3 pairs with R4, or (2) R2 pairs with R3. The region R1 contains 2 adjacent Trp codons (UGG). The pairing of R3 and R4 generates a GC palindrome followed by 8 successive U residues, a typical termination signal for transcription. This sequence forming the termination stem-loop is an attenuator (Fig. 5.8).

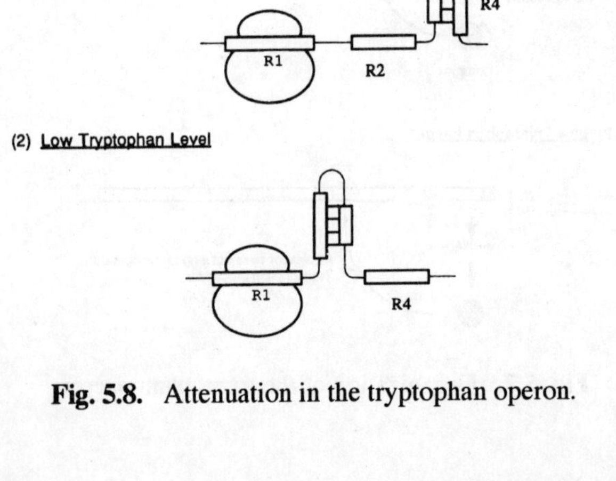

Fig. 5.8. Attenuation in the tryptophan operon.

At a low concentration of tryptophan, the ribosome translating the leader sequence proceeds until it comes to the Trp (UGG) codons. The process will stall, because there is a lack of Trp-tRNA for translation. The stalling enables R2 to pair with R3 thus preventing the stem-loop terminator formed between R3 and R4. Under this condition, the *trp* genes are transcribed and translated to the respective enzymes for the biosynthesis of the tryptophan amino acid.

At a high concentration of trptophan, Trp-tRNA is present in abundance, and the function of ribosome proceeds smoothly. Under this condition, the R3R4 termination stem-loop is formed. Transcription terminates at the end of the leader sequence.

5.4.3 Hybrid Promoters

Functional hybrid promoters can be designed to possess desirable properties. For example, *tac* promoter, a strong promoter frequently used in bacterial systems, is such a hybrid. The -35 region is derived from the *trp* promoter, and the Pribnow box (-10 region) from the *lac* promoter. The *tac* promoter is more efficient than the two parent promoters, and is useful for controlled expression of foreign genes at high levels in *E. coli.* The *tac* promoter is controlled by the *lac* repressor and can be derepressed by IPTG (see Section 9.1.1).

5.5 The Control System in Eukaryotic Cells

The transcription/translation process described thus far applies generally to prokaryotes, such as *E. coli*. In higher organisms, several additional features are crucial to the understanding of gene cloning.

5.5.1 Transcriptional Control

Analogous to bacterial transcriptional control, there are conserved sequences in eukaryotic promoters, located at -25 to -35 region, called the TATA box (or Hogness box). In addition, there are two frequently found conserved short sequences, the GC box and CAAT box. These are called enhancers that are involved in the activation of transcription. There are also negatively acting DNA sequences- called silensers- which are involved in the repression of transcription. The location of enhancers and silensers can be upstream or downstream of the promoter, or even thousands of bases away from the promoter.

In contrast to bacterial system, there are 3 distinct RNA polymerases in the synthesis of eukaryotic RNA - RNA polymerase I for the synthesis of rRNA, RNA polymerase II for mRNA synthesis, and RNA polymerase III for tRNA. Eukaryotic RNA polymerase II acts in cooperation with a number of proteins, called transcription factors, TFIIA, B, C, D, E, F and H. In the initiation of transcription, TFIID binds to the TATA box, followed by other transcription factors associating with RNA polymerase II to form an initiation complex. The complex formation allows correct positioning of the enzyme at the transcription start site, unwinding of the DNA at the site, and the enzyme proceeding from the promoter onto the encoding gene sequence. In this regard, the function of an eukaryotic initiation complex is analogous to the bacterial RNA polymerase holoenzyme.

Enhancers and silencers are docking sites for a group of transcription factors that includes zinc finger proteins, leucine zipper proteins, and helix-turn-helix proteins. Each protein class interacts with DNA in a different manner. For example, leucine zipper proteins are characterized by a repeat of leucine residues called zipper region. A dimer is formed by the association of two leucine zipper regions (Fig. 5.9). The pairing of zipper proteins produces dimers of various combinations to effect different responses in DNA interactions.

The assembly of promoter, enhancer (silencer), and transcription factors constitute the molecular machinery for controlling the transcriptional activity of a gene. The specific combination of transcription factors and their interactions with DNA sequences plays a central role in which genes are differentially expressed, giving rise to specific cell types.

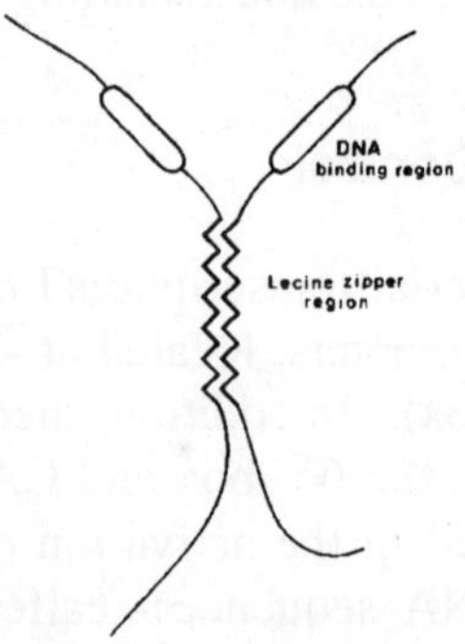

Fig. 5.9. Leucine zipper association to form a dimer.

5.5.2 Introns and Exons

In eukaryotes, the coding region of mRNA is interrupted by segments of DNA that do not encode amino acids. These noncoding DNA segments are called introns (Fig. 5.10). The coding sequences are exons. The introns are removed before translation in a process called splicing. There are 4 types of introns, each with their own special features and a different splicing mechanism. The "GC-AT" class of introns is the most commonly encountered, and the more thoroughly studied. These introns have the dinucleotides GC and AT at their 5' and 3' ends. The full consensus sequences are 5'-AGGTAA^GT at the 5' splice site, and 5'-$(Py)_6$NCAG^ at the 3' splice site (Py = C or U, and N = any nucleotide). Cleavage occurs at the 5' splice site, for a free 5' end which attaches to an internal site in the intron to form a lariat structure. The 3' splice site is then cleaved and the two exons are joined together. Some genes may have the mRNA spliced in more than one way. Therefore depending on the splicing, different forms of a protein can be produced. It is suggested that each human gene can on average spell out three proteins by using different combinations of exons.

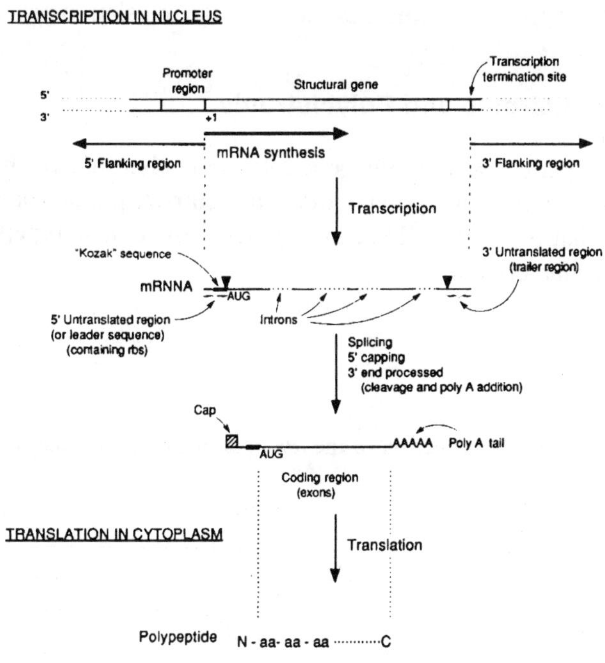

Fig. 5.10. General scheme of transcription and translation in eukaryotic cells.

5.5.3 Capping and Tailing

Following transcription, the 5' end of a mRNA is capped by the addition of a methylated guanosine nucleotide (Fig. 5.10). The 7-methylguanine (m7G) is added to the 5' phosphate end of the mRNA after transcription by a two-step process: attachment of the G to the phosphate followed by methylation of the G base nitrogen at position 7. The cap structure may play a role in the translation process, but its definite function is still unclear.

The 3' end of the mRNA is processed by the addition of ~20-200 adenosine (poly A) tail. This process occurs between 10-30 bases down-stream of a specific polyadenylation signal with a consensus sequence 5'-AAUAAA. Polyadenylation may increase stability of the mRNA, and is effectively the termination process for RNA polymerase II transcription.

5.5.4 Ribosome Binding Sequence

Instead of the Shine-Dalgarno sequence in prokaryotic mRNAs, eukaryotic ribosomes utilizes efficiently a sequence in the mRNA known as the Kozak sequence: GCCGCCACC<u>AUG</u>G, which lies within a short 5' untranslated region, for binding and initiating the translation process.

5.5.5 Monocistronic and Polycistronic

Eukaryotic mRNAs are generally monocistronic in that a single mRNA translates into one polypeptide. In contrast, prokaryotes are poly-cistronic in that a single mRNA may produce several polypeptides as in the case of the *lac* operon.

Review

1. List and describe the function of the *lac* operon strucural genes and regulatory elements.

Structural genes	Functions
(A) ______________	____________________
(B) ______________	____________________
(C) ______________	____________________
Regulatory elements	**Functions**
lac promoter	____________________
lac repressor gene	____________________
lac operator	____________________

2. Describe the effects of the presence and absence of lactose on the following control elements in the *lac* operon.

	Lactose present	Lactose absent
(A) *lac* promoter		
(B) *lac* repressor protein		
(C) *lac* operator		

3. How does RNA polymerase recognize the start site and termination site in transcription? What is the function of sigma factors?
4. What is the functional role of a ribosome binding site? How is it related to the start codon AUG in mRNA? What is the termination site for translation?
5. In the *lac* operon, lactose acts as an inducer. In gene cloning, the genetic switch for the *lac* operon is turned on by IPTG. Why is IPTG used instead of lactose?
6. In the *lac* operon, glucose is a suppressor. Describe the sequence of events occurred when glucose is added to the growth medium.
7. Describe the functions of the *trp* operon regulatory elements.
   (A) *trp* promoter
   (B) *trp* operator
   (C) Repressor gene (*trpR*)
   (D) Leader (*trpL*)
8. Describe the effects of intracellular tryptophan concentration.

	High Trp level	Low Trp level
(A) Repressor protein		
(B) *trp* promoter		
(C) *trp* operator		

9. Explain the mechanism of attenuation using the *trp* operon as an example.
10. What are the components of the *tac* promoter?
11. Describe the functions of the following transcriptional regulatory elements in eukaryoic cells: (A) enhancer, (B) transcription factor.
12. What is the function of a Kozak sequence?

CHAPTER 6

READING THE NUCLEOTIDE SEQUENCE OF A GENE

In this Chapter, we will apply what have been learned in the previous chapters to read the nucleotide sequences of a prokaryotic and a eukaryotic gene. The aim is to provide a simple guide for reading a gene sequence. Starting with a DNA sequence, one can extract a wealth of information on the architectural organization of the gene, including many of the features at the protein level. Both transcriptional and translation processes can be inferred by reading a gene sequence (Refer to previous chapters for basic concepts).

6.1 The *E. coli dut* gene

For a prokaryotic gene, we use the *E. coli dut* gene as an example (Fig. 6.1). The dut gene codes for the enzyme dUTPase (deoxyuridine 5'-triphosphate nucleotidohydrolase, E.C. 3.6.1.23) (Lundberg et al. 1983. *EMBO J.* 2, 967-971).

dUTPase is a phosphatase that removes diphosphates from dUTP, a reaction in pyrimidine metabolism (dUTP + H_2O = dUMP + diphosphate). The enzyme is ubiquitously found in both prokaryotes and eukaryotes.

`Notice that by convention, the gene is named by three italicized letters in the lower case, while the protein name in abbreviation is in normal type in capital letters. E.C. 3.6.1.23 is the enzyme code number specifically for dUTPase, according to the Nomenclature Committee of the International Union of Biochemistry and Molecular Biology (see Section 3.5 for Enzyme Classification.)

Let us take a detailed look at the *E. coli* dUTPase nucleotide sequence. For simplicity, features to be discussed are highlighted, and some are labeled. The deduced amino acids are presented below the DNA sequence, corresponding to their respective codons.

```
CAGAGAAAATCAAAAAGCAGGCCACGCAGGGTGATGAATTAACAATAAAAATGGTTAAAA   60
ACCCCGATATCGTCGCAGGCGTTGCCGCACTAAAAGACCATCGACCCTACGTCGTTGGAT  120
TTGCCGCCGAAACAAATAATGTGGAAGAATACGCCCGGCAAAAACGTATCCGTAAAAACC  180
TTGATCTGATCTGCGCGAACGATGTTTCCCAGCCAACTCAAGGATTTAACAGCGACAACA  240
                                                 -35 region
ACGCATTACACCTTTTCTGGCAGGACGGAGATAAAGTCTTACCGCTTGAGCGCAAAGAGC  300

          Pribnow box    +1          rbs          +1
TCCTTGGCCAATTATTACTCGACGAGATCGTGACCCGTTATGATGAAAAAAATCGACGTT  360
                                                M  K  K  I  D  V
AAGATTCTGGACCCGCGCGTTGGGAAGGAATTTCCGCTCCCGACTTATGCCACCTCTGGC  420
 K  I  L  D  P  R  V  G  K  E  F  P  L  P  T  Y  A  T  S  G
TCTGCCGGACTTGACCTGCGTGCCTGTCTCAACGACGCCGTAGAACTGGCTCCGGGTGAC  480
 S  A  G  L  D  L  R  A  C  L  N  D  A  V  E  L  A  P  G  D
ACTACGCTGGTTCCGACCGGGCTGGCGATTCATATTGCCGATCCTTCACTGGCGGCAATG  540
 T  T  L  V  P  T  G  L  A  I  H  I  A  D  P  S  L  A  A  M
ATGCTGCCGCGCTCCGGATTGGGACATAAGCACGGTATCGTGCTTGGTAACCTGGTAGGA  600
 M  L  P  R  S  G  L  G  H  K  H  G  I  V  L  G  N  L  V  G
TTGATCGATTCTGACTATCAGGGCCAGTTGATGATTTCCGTGTGGAACCGTGGTCAGGAC  660
 L  I  D  S  D  Y  Q  G  Q  L  M  I  S  V  W  N  R  G  Q  D
AGCTTCACCATTCAACCTGGCGAACGCATCGCCCAGATGATTTTTGTTCCGGTAGTACAG  720
 S  F  T  I  Q  P  G  E  R  I  A  Q  M  I  F  V  P  V  V  Q
GCTGAATTTAATCTGGTGGAAGATTTCGACGCCACCGACCGCGGTGAAGGCGGCTTTGGT  780
 A  E  F  N  L  V  E  D  F  D  A  T  D  R  G  E  G  G  F  G
CACTCTGGTCGTCAGTAACACATACGCATCCGAATAACGTCATAACATAGCCGCAAACAT  840
 H  S  G  R  Q Stop                 Stop                  Stem-

TTCGTTTGCGGTCATAGCGTGGGTGCCGCCTGGCAAGTGCTTATTTTCAGGGGTATTTTG  900
loop                           Second stem-loop

TAACATGGCAGAAAAACAAACTGCGAAAAGGAACCGTCGCGAGGAAATACTTCAGTCTCT  960
GGCGCTGATGCTGGAATCCAGCGATGGAAGCCAACGTATCACGACGGCAAAACTGGCCGC 1020
CTCTGTCGGCGTTTCCGAAGCGGCACTGTATCGCCACTTCCCCAGTAAGACCCGCATGTT 1080
CGATAGCCTGATTGAGTTTATCGAAGATAGCCTGATTACTCGCATCAACCTGATTCTGAA 1140
AGATGAGAAAGACACCACAGCGCGCCTGCGTCTGATTGTGTTGCTGCTTCTCGGTTTTGG 1200
TGAGCGTAATCCTGGCCTGACCCGCATCCTCACTGGTCATGCGCTAATGTTTGAACAGGA 1260
TCGCCTGCAAGGGCGCATCAACCAGCTGTTCGAGCGTATTGAAGCGCAGCTGCGCCAGGT 1320
ATTGCGTGAAAAGAGAATGCGTGAGGGTGAAGGTTACACCACCGATGAAACCCTGCTGGC 1380
AAGCCAGATCCTGGCCTTCTGTGAAGGTATGCTGTCACGTTTTGTCCGCAGCGAATTTAA 1440
ATACCGCCCGACGGATGATTTTGACGCCCGCTGGCCGCTAATTGCGGCCAGTTGCAGTAA 1500
TATGACGCCGGATGACTTTTCATCCGGCGAGTTTCTTTAAACGCCAAACTCTTCGCGATA 1560
GGCCTTAACCGCCGCCAGATGTTCCGCCATTTCCGGCTTCTCTTCCAGG 1609
```

Fig. 6.1. DNA sequence of the *E. coli dut* gene (Lundberg et al. 1983. *EMBO J.* 2, 967-971). The deduced amino acid sequence is shown below the DNA sequence. Special regions of interests are in bold, underlined, labeled, and explained in details in the text.

(1) The nucleotide sequence consists of 1609 bp, and is labeled 1 to 1609. The sequence corresponds to part of a genomic DNA fragment isolated from the genomic library of *E. coli* K-2. This DNA fragment carries the *dut* gene region plus extended upstream and downstream sequences.

(2) The sequence by convention is written in 5'->3' direction of the coding strand (same sequence as the mRNA).

(3) The nucleotide sequence contains several identifiable *transcriptional* elements.

    (a) Promoter –35 region (286-291, TTGAGC)
    (b) Pribnow box –10 region (310-316, AATTATT)
    (c) Transcription start site (323)
    (d) Transcription termination site (831-851, stem-loop structure)

    During transcription, the RNApolymerase recognizes the promoter region, and starts transcription at position 323. This transcription start site is the +1 nucleotide. The upstream sequence is numbered minus, hence –10 region and –35 region in the promoter.

    The mRNA reads from the +1 nucleotide (between the Pribnow box and the ribosome binding site, rbs) and extends to the stem-loop region. There is also a second stem-loop from positions 866 to 893, with imperfect pairing at one of the base pairs. This second stem-loop is GC-rich and has a number of Ts following it, a common feature found in many bacterial transcription terminators.

(4) The nucleotide sequence contains several *translational* elements.

    (a) Ribosome binding site (Shine-Dalgarno sequence, 330-333 GTGA)
    (b) Translation start codon (343-345, ATG)
    (c) Translation stop codon (796-798, TAA)

    The Shine-Dalgarno sequence is to position the ribosome for translation. The ATG start codon is the +1 nucleotide when referring to the translation process. There is a second ATG codon at position 340, but it has been determined experimentally that this is not the start signal for translation.

    Translation of the mRNA starts at the ATG start codon and extends to the codon CAG immediately preceding the stop codon. This nucleotide sequence is the open reading frame corresponding to the structural gene sequence of dUTPase. The resulting polypeptide chain consists of 150 amino acids, beginning with a methionine and ends with a glutamine. The dTUPase protein has a calculated molecular weight of 16,006.

(5) There is a second open reading frame downstream of the *dut* gene, starting with the ATG at position 905 and ends with TAA at position

1538, coding for another unknown protein. This open reading frame sequence may be co-transcribed with *dut* gene. It is common that prokaryotic genes are polycistronic, having more than one structural gene under the control of the same promoter.

6.2 The Human *bgn* gene

Let us proceed to examine the nucleotide sequence of the human biglycan gene, which exemplifies the more complex architecture of eukaryotic systems. Biglycan is a small leucine-rich proteoglycan found ubiquitously in the pericellular matrix of a variety of cells, and plays an important role in connective tissue metabolism. The protein contains two attached glycosaminoglycan chains, and is thus called biglycan for this reason.

6.2.1. Reading the genomic sequence

We start first with reading the genomic DNA sequence of the human biglycan gene (Fisher, et al. 1991. *J. Biol. Chem.* 266, 14371-14377). The sequence is presented in Fig. 6.2.

(1) Biglycan expression in the physiological state is highly regulated by transcription factors and other feedback mechanisms. All sequences relating to the regulatory elements are located in the 5' flanking region where the functional promoter activity resides.

AP-1	transcription factor AP-1
AP-3	transcription factor AP-3
IL-6RE	interleukin-6 response element
TGF-β	transforming growth factor-β
TGF-βNE	TGF-negative element
TNF-α	tumor necrosis factor
GRE	glucocorticoid response element

(2) The biglycan gene promoter lacks both CAAT and TATA boxes, but is rich in GC content in two enriched regions, -1 to -164 (73%) and -204 to -256 (87%).

(3) The genomic gene consists of 8 exons and 7 introns. The entire nucleotide sequence is ~8 kb in length. Transcription starts from the +1 nucleotide encompassing all exons and introns. The latter sequences are then removed by RNA splicing. Exon I is located entirely in the untranslated 5' sequence. Translation occurs from the ATG at position 1344 in exon II and extends to exon III to VIII.

(4) The cDNA gene of biglycan isolated from the cDNA library would give the exact sequence of the exons linked together after splicing. Notice that some nucleotides at the exon-intron junctions are removed during the splicing process.

```
         IL6-RE                              IL6-RE     IL6-RE
GAGCTCCCCT GGGAGCATCC TCCCTGGCCT GGGACCTCCC AGACCCCACC
CCCCGGTTGA GTGATGGCAC TGCCAGGGGT TGAAGACCCT CAGCCCTCGA -1119
CGTTGTCCTC TCTCCATTGG ATGCCGCCTC TCTCTAGCCA CCCCTCTCTC
CCTCTCTGCC CCTTCGAGCT TTTTCTCTCA ATATGCAATT TTCTCTTTTG -1019
GTCTTCCGCA CTCTTGGCCC CCAGTTCTAT TGCAGATCTG TTTCTCACTC
CATCTAAACT CTTACCCCTG TGTCTCAGGA GCTGCTCTTG CTGAGGGAAG  -919
AAGGGGACAC TACGGGACAG GGGGGCAGTG TCGTACTAAG GACCTGGGCT
CTAGCCACTG GAGGAACTGG ACTCATTTGG GCCCTCAGGA AGCGGCTGAG  -819
  TGF-βNE
TCTTGGTGGG GTAACCCGGT TAGCCCCCGT AAGTGACCAG CACAGGGCTG
                 AP-2
AGCCCAGAGG AAGTGGCCAC CCACAGAGTG GTTCTCATGT CCGAGGGGAC  -719
                         GRE                      AP-1
CTGCAGGGAT TGAGCAAGAA GACTGACTCG CTGGATCCTT CGTCTCTGAA
TCAGTTCAGG GCAGGCAAGC TGGGGAGCCC CCTGCCCCGT CCTGCCACCA  -619
CCAGCCGGAT CGGGCCCTCT TTTTAAGGGA AGAAAGTCTG AAGTGGAAGG
                                  IL6-RE     AP-3
GAGGGCACAG GGCGCCAGGA GCCTACATGA AGTCCTTCCA GAAATCCACA  -519
ACAGCTACCT CTCTGATCCT GGAGAAACCA CCTCCTTGCT TAGGCCCAAG
CAGGTTCCTG GCAGGCTCAG GACCAAATTC CAGGGGCCAC TCATGGGCCT  -419
AGCAGCCCAA GGCCGCCTCC CCCTCGTCTT TCTTCCATCT CTCTTTCCTC
TGCCTGGCGA GATGCCAGCC AGCACCTCAG TGTCCCCATC TGGGCAGTGG  -319
      GRE
AAAGTTTGAC TCTCTGGGTC CTTGTTTGAG TGAGTGCGAG TGTGTCCGTT
                    AP-2
CCTTTGCTGT CTGCCCCAGG CGGGGGAGGG GGGGGGAGGT GGTGGGGGCG  -219
  SP1
AGGGGGCGGG GGCTCAGCTA GTCCAGCCGT CTACAAGAAA ATTGCTCCCT
                   IL6-RE                         SP1
TTGAAGCTGC CAGGGGGGCC GGGAAGCCTG CCCCCTCCTG CTCGCCCGCC  -119
    SP1                                   SP1
CTCTCCGCCC CACCAGCCCC CTCCCTCCTT TCCTCCCTCC CCGCCCTCTC
                                                  SP1
CCCGCTGTCC CCTCCCCGTC GGCCCGCCTG CCCAGCCTTT AGCCTCCCGC   -19
                  +1[exon I  >
CCGCCGCCTC TGTCTCCCTC TCTCCACAAA CTGCCCAGGA GTGAGTAGCT    32
GCTTTCGGTC CGCCGGACAC ACCGGACAGA TAGACGTGCG GACGGCCCAC
CACCCCAGCC CGCCAACTAG TCAGCCTGCG CCTGGCGCCT CCCCTCTCCA   132
GGTAGGGCTG GCTTCAAGCT GCCTCCTCAG CAACCCAGAG ATGCCCCTGG
CTCTGCTGCC TCCGCTGTCC CAAGCCCTGG TCCTGCTGTC CCCAGTGCCG   232
CGAGGGTGTC CACAGATTTC CCCGGTGCTC TCTGTAGGCT GCTGATCCAC
GCCCCTTCAT CGCCACCCTG CGGCCCCCTT GGTCCCTGTC AGGCTTCTGC   332
TCGTCTCGCC GCCCTCCAGG CACCTTTCCC TCACCCCTTC CTCTCCCTTC
TGACCTTGCT CTGCTTCATC CACCTCTTGT CTCTCTGCCT CCCACTCGGG   432
GTCCGTCTTC TTGGCTACCA CCCTAGAGCG TGGCTGGGTG ACTGGTACCC
CAGCTTTGCC AATGGCCCTG TTTCATCATT GCAAGTCCCA GGCGCATGCT   532
CCACTCCCTC AGCCTCGCTC TGCCCAGGCG CCTCCTTGCT CCAGGCTTGG
CGCCTGGCCC GGGTTGGGTC GGATCGGGGA GGACCGCCCA GCGCCCACCG   632
```

AGCTC..............650bp................. ACAGGTGGGT GCTGGTGCTG ATGATCCCCT

[exon II >

CGCCTCTTCC CCCAG**GTCCA TCCGCCATGT GGCCCCTGTG GCGCCTCGTG** 1367

M W P L W R L V

TCTCTGCTGG CCCTGAGCCA GGCCCTGCCC TTTGAGCAGA GAGGCTTCTG

S L L A L S Q A L P F E Q R G F W

GGACTTCACC CTGGACGATG GGCCATTCAT GATGAACGAT GAGGAAGCTT 1467

D F T L D D G P F M M N D E E A

CGGGCGCTGA CACCTCGGGC GTCCTGGACC CGGACTCTGT CACACCCACC

S G A D T S G V L D P D S V T P T

TACAGCGCCA TGTGTCCTTT CGGCTGCCAC TGCCACCTGC GGGTGGTTCA 1567

Y S A M C P F G C H C H L R V V Q

GTGCTCCGAC CTGGGTTTGT CCCTGAGTGA TGGGGAGCGG GGCATGCAGG

C S D L

GAGGCTCAGG TGCAGCCTGA GAGCCCCTTC TGAAGGGGGC ACATGCTGGT 1667

CCTGTGGACG GTGGCGAGCA TGATGTAAGT GTAGGAGGGG TCCAGCCGTC

TGGCTGTGAG CTGTGCAGTT TGTGCCCACT TGTGGTGGCA TCCCCGTGTG 1767

CCCGTCAGTG TCCCTGTGTG TGTGTCCCCG GTCCTCCCTA CCAGTGGGGC

TAGTCGGCTG GATGGCTCCA AGTTCATGCT GGTGATGGTG GTGGGGCCCC 1867

[exon III >

TAG**GTCTCGA GTTCATGCTG GTGGTGGGGG TGGGGCCCCT AGGTCTCAAG**

TTCATGCTGG TGATGGGGGT GGGGCCCCTA GGTCTGAAGT CTGTGCCCAA 1967

G L K S V P K

AGAGATCTCC CCTGACACCA CGCTGCTGGA CCTGCAGAAC AACGACATCT

E I S P D T T L L D L Q N N D I

CCGAGCTCCG CAAGGATGAC TTCAAGGGTC TCCAGCACCT CTACGTAAGG 2067

S E L R K D D F K G L Q H L Y

AGCTGGGAGG AACCAGCAGG CCTACAGCAG AGGGCAGGGG TCCGGGTGGG

TGCATGTGCG TGGACGTGTG GGGTATGAGA GGGGTTCGGG GACTCGTGGG 2167

ACTTCAGGGT GAAGCCTGGA GCCAGCCGTG ATGGGAGCTC CCGGGTTTGC

GGCTCACTCA TGTGGGTTTG AGCAACCACA GCTGCAGGAC CGGATCGCTC 2267

AGTTCGGCTC CCTTCGTGGC TGAAAACGTT TCATCACGTC CACTCCTCCC

AGCAACAGAG GAGAACGGAT TTCATTGTAG CCAGTGTGCG TGTGAGGAAA 2367

CTGAGGCTGG GAGCGGCAAG GCAGTGGTGG CACTGCTGGG GCTCAGGACC

GGGCCTGGGT GCTGCCTCCT GCCCTGCACT CTGCTCACAA GCATGGACTG 2467

ACCTCCTCGA GCGCCAGTGG GCTGGGGAGG CACAGGAAGG CAGGAGAGAG

GGGCGGGTGG GGTGGGGAGT CTGTGCCTTC ACCTCCTCCG CCCACCCTGC 2567

[exon IV >

TTCAG**GCCCT CGTCCTGGTG AACAACAAGA TCTCCAAGAT CCATGAGAAG**

A L V L V N N K I S K I H E K

GCCTTCAGCC CACTGCGGAA CGTGCAGAAG CTCTACATCT CCAAGAACCA 2667

A F S P L R N V Q K L Y I S K N H

CCTGGTGGAG ATCCCGCCCA ACCTACCCAG CTCCCTGGTG GAGCTCCGCA

L V E I P P N L P S S L V E L R

TCCACGACAA CCGCATCCGC AAGGTGCCCA AGGGAGTGTT CAGTGGGCTC 2767

I H D N R I R K V P K G V F S G L

```
CGGAACATGA ACTGCATCGG TGAGCTGAGG GCCTCCCAGA ACATTCCAGA
 R  N  M    N  C  I
GCCTTGTCTC GAGGCATGGG GAAGGGAGAC CAAGGAATAC CTTTAGAGGC  2867
TCAGTTCAAG AAAGAGTATG GTGAGAACGG TCAAAAGAAA ATCCATGGAT
TTCTTGGCAA ATCCTCCATG CAGGCGATCA CCACGGCTAA AGAGAAGACT  2967
GGCCAGAGGG GCCGGGTGGC TTCCGGAGCC CCATCTTCAT CTCTGGCACT
CCTCCCTTTC CTCTTGCTGC CCCTGGAGCT AGCAGTCCTG GGGCTAGCAG  3067
TCCTGAACAG CTAGGAGTTT GCAATTAGCC CGGTAAATTA GCAGAACTGC
TTTCAGGAGA CGGGAGCAGC CGGCAGGTAG CAGGGCCCAC CACACTGGCC  3167
CGGAAGTGAC AGGACCCAGG GCTGTGCAGG GACCACCAGG CTCCCGGGCT
                    [exon V  >
AATGAGGTCT CTCCCCTAGA GATGGGCGGG AACCCACTGG AGAACAGTGG  3267
                       M  G  G   N  P  L   E  N  S  G
CTTTGAACCT GGAGCCTTCG ATGGCCTGAA GCTCAACTAC CTGCGCATCT
  F  E  P   G  A  F   D  G  L  K   L  N  Y   L  R  I
CAGAGGCCAA GCTGACTGGC ATCCCCAAAG GTAGGAAGCC CACTCTTCCT  3367
S  E  A  K   L  T  G   I  P  K
GCACGCCTGC CTGCCTCACC CCCAACAGCA CAGATGGCCA GGGTGGGGGC
TCTGGATGGG CCCGATCTAC TCAGGGAAAG GCTCAACAGT CCCCTCCCGC  3467
CACCTGGGGC AGAGCTAGGG CCCCTGCCCT CAGCACCTGC ATTCTCCCCT
                      [exon VI  >
GTGCCCTCTT CTCCTGGCAG ACCTCCCTGA GACCCTGAAT GAACTCCACC  3567
                        L  P  E   T  L  N   E  L  H
TAGACCACAA CAAAATCCAG GCCATCGAAC TGGAGGACCT GCTTCGCTAC
L  D  H  N   K  I  Q   A  I  E   L  E  D  L   L  R  Y
TCCAAGCTGT ACAGGTGAGG CCAGCAGGGC ACCGCCAAGG GTGATGCCAG  3667
 S  K  L   Y
AGTCCCTCAG TGCTGTGTGG CCCCTCGCGC CCAGCCCCCC ATCCTTACCT
                                     [exon VII  >
CCAGCCTTTG AGTCCGTGTC ATTCTCCCGC TCACAGGCTG GGCCTAGGCC  3767
                                        L   G  L  G
ACAACCAGAT CAGGATGATC GAGAACGGGA GCCTGAGCTT CCTGCCCACC
H  N  Q  I   R  M  I   E  N  G   S  L  S  F   L  P  T
CTCCGGGAGC TCCACTTGGA CAACAACAAG TTGGCCAGGG TGCCCTCAGG  3867
 L  R  E   L  H  L  D   N  N  K   L  A  R   V  P  S  G
GCTCCCAGAC CTCAAGCTCC TCCAGGTGAG AGCTGGGCAT GCACAGCCAG
 L  P  D   L  K  L   L  Q

G......................1200bp...................... ACCTCACACC ACCAAACACA CCTCTACCCC  5148
AGCCCCGCCC CCACATGTCC TCAACCTGAC CCACCTGAGA CCCTCATCCT
TGTCCCTGGT CACATCCAGT GCCTTAATCC TGGCTGACAC CCACACAAAT  5248
AACACGCCCA TGCCTTGGTT TGCTCCTCCC AACAACGGGG AGCCTCTGGT
GTGGCCCTTG AAGTAGGTTG CAGAGGCAAC AGCAAAATGC CTCCTGGAGG  5348
CAGCGGGCTT GGCGTGGAGG GAGGGAGGCC TGTGACCCGG CCTCTCTGCC
    [exon VIII  >
TTCAGGTGGT CTATCTGCAC TCCAACAACA TCACCAAAGT GGGTGTCAAC  5448
      V  V   Y  L  H   S  N  N   I  T  K  V   G  V  N
GACTTCTGTC CCATGGGCTT CGGGGTGAAG CGGGCCTACT ACAACGGCAT
 D  F  C   P  M  G  F   G  V  K   R  A  Y   Y  N  G  I
```

```
CAGCCTCTTC AACAACCCCG TGCCCTACTG GGAGGTGCAG CCGGCCACTT  5548
  S  L  F    N  N  P    V  P  Y  W    E  V  Q    P  A  T
TCCGCTGCGT CACTGACCGC CTGGCCATCC AGTTTGGCAA CTACAAAAAG
F  R  C  V    T  D  R    L  A  I    Q  F  G  N    Y  K  K
TAGAGGCAGC TGCAGCCACC GCGGGGCCTC AGTGGGGGTC TCTGGGGAAC  5648
ACAGCCAGAC ATCCTGATGG GGAGGCAGAG CCAGGAAGCT AAGCCAGGGC
CCAGCTGCGT CCAACCCAGC CCCCCACCTC AGGTCCCTGA CCCCAGCTCG  5748
ATGCCCCATC ACCGCCTCTC CCTGGCTCCC AAGGGTGCAG GTGGGCGCAA
GGCCCGGCCC CCATCACATG TTCCCTTGGC CTCAGAGCTG CCCCTGCTCT  5848
CCCACCACAG CCACCCAGAG GCACCCCATG AAGCTTTTTT CTCGTTCACT
CCCAAACCCA AGTGTCCAAA GCTCCAGTCC TAGGAGAACA GTCCCTGGGT  5948
CAGCAGCCAG GAGGCGGTCC ATAAGAATGG GGACAGTGGG CTCTGCCAGG
GCTGCCGCAC CTGTCCAGAA CAACATGTTC TGTTCCTCCT CCTCATGCAT  6048
TTCCAGCCTT G.........1300bp......... GGACAGCGGT CTCCCCAGCC
TGCCCTGCTC AGCCCTGCCC CCAAACCTGT ACTGTCCCGG AGGAGGTTGG  7429
GAGGTGGAGG CCCAGCATCC CGCGCAGATG ACACCATCAA CCGCCAGAGT
CCCAGACACC GGTTTTCCTA GAAGCCCCTC ACCCCCACTG GCCCACTGGT  7529
GGCTAGGTCT CCCCTTACTC TTCTGGTCCA GCGCAACCAG GGGCTGCTTC
TGAGGTCGGT GGCTGTCTTT CCATTAAAGA AACACCGTGC             7619
```

Fig. 6.2. Genomic DNA sequence of the biglycan gene (Fisher et al. 1991. *J. Biol. Chem.* 266, 14371-14377. Special regions of interests are in bold, underlined, labeled, and explained in details in the text.

6.2.2 Reading the cDNA sequence

Let us turn our attention to the sequence of the biglycan cDNA gene presented in Fig. 6.3. Again, important features are highlighted, labeled, or underlined to facilitate the discussion in the text.

(1) First, notice that the sequence is labeled "*Homo sapiens* (human) biglycan mRNA" in the legend as obtained from the GenBank database (www.ncbi.nlm.nih.gov/entrez/ Accession number BC002416). The sequence is actually presented as the cDNA sequence. It is understood that the coding sequence corresponds to the mRNA sequence, except for the U to T base replacement.

(2) The mRNA contains the first 172 bp sequence as the 5' untranslated region. It has a polyA signal (position 2357-2362) and a polyA tail (position 2371-2401) at the 3' end. Although not shown, the 5' end is capped as in all eukaryotic mRNA. Notice that downstream from the 3' end of the gene sequence is a ~1100 bp extension (position 1267 to 2401). A long trailer region like this is typically found in eukaryotic mRNA.

AGCCTCCCGC CCGCCGCCTC TGTCTCCCTC TCTCCACAAA CTGCCCAGGA
GTGAGTAGCT GCTTTCGGTC CGCCGGACAC ACCGGACAGA TAGACGTGCG 100
GACGGCCCAC CACCCCAGCC CGCCAACTAG TCAGCCTGCG CCTGGCGCCT

Start codon
CCCCTCTCCA GGTCCATCCG CC**ATG**TGGCC CCTGTGGCGC CTCGTGTCTC 200
M W P L W R L V S
(Signal peptide)

TGCTGGCCCT GAGCCAGGCC *CTGCCCTTTG AGCAGAGAGG CTTCTGGGAC*
L L A L S Q A *L P F E Q R G F W D*
Proprotein >

*TTCACCCTGG ACGATGGGCC ATTCATGATG AAC***GATGAGG AAGCTTCGGG** 300
F T L D D G P F M M N **D E E A S G**
Mature protein >

CGCTGACACC TCGGGCGTCC TGGACCCGGA CTCTGTCACA CCCACCTACA
A D T S G V L D P D S V T P T Y
GCGCCATGTG TCCTTTCGGC TGCCACTGCC ACCTGCGGGT GGTTCAGTGC 400
S A M C P F G C H C H L R V V Q C
TCCGACCTGG GTCTGAAGTC TGTGCCCAAA GAGATCTCCC CTGACACCAC
S D L G L K S V P K E I S P D T T
GCTGCTGGAC CTGCAGAACA ACGACATCTC CGAGCTCCGC AAGGATGACT 500
L L D L Q N N D I S E L R K D D
TCAAGGGTCT CCAGCACCTC TACGCCCTCG TCCTGGTGAA CAACAAGATC
F K G L Q H L Y A L V L V N N K I
TCCAAGATCC ATGAGAAGGC CTTCAGCCCA CTGCGGAAGC TGCAGAAGCT 600
S K I H E K A F S P L R K L Q K L
CTACATCTCC AAGAACCACC TGGTGGAGAT CCCGCCCAAC CTACCCAGCT
Y I S K N H L V E I P P N L P S
CCCTGGTGGA GCTCCGCATC CACGACAACC GCATCCGCAA GGTGCCCAAG 700
S L V E L R I H D N R I R K V P K
GGAGTGTTCA GCGGGCTCCG GAACATGAAC TGCATCGAGA TGGGCGGGAA
G V F S G L R N M N C I E M G G N
CCCACTGGAG AACAGTGGCT TTGAACCTGG AGCCTTCGAT GGCCTGAAGC 800
P L E N S G F E P G A F D G L K
TCAACTACCT GCGCATCTCA GAGGCCAAGC TGACTGGCAT CCCCAAAGAC
L N Y L R I S E A K L T G I P K D
CTCCCTGAGA CCCTGAATGA ACTCCACCTA GACCACAACA AAATCCAGGC 900
L P E T L N E L H L D H N K I Q A
CATCGAACTG GAGGACCTGC TTCGCTACTC CAAGCTGTAC AGGCTGGGCC
I E L E D L L R Y S K L Y R L G
TAGGCCACAA CCAGATCAGG ATGATCGAGA ACGGGAGCCT GAGCTTCCTG 1000
L G H N Q I R M I E N G S L S F L
CCCACCCTCC GGGAGCTCCA CTTGGACAAC AACAAGTTGG CCAGGGTGCC
P T L R E L H L D N N K L A R V P
CTCAGGGCTC CCAGACCTCA AGCTCCTCCA GGTGGTCTAT CTGCACTCCA 1100
S G L P D L K L L Q V V Y L H S
ACAACATCAC CAAAGTGGGT GTCAACGACT TCTGTCCCAT GGGCTTCGGG
N N I T K V G V N D F C P M G F G

```
GTGAAGCGGG CCTACTACAA CGGCATCAGC CTCTTCAACA ACCCCGTGCC  1200
 V  K  R   A  Y  Y  N   G  I  S   L  F  N   N  P  V  P
CTACTGGGAG GTGCAGCCGG CCACTTTCCG CTGCGTCACT GACCGCCTGG
 Y  W  E   V  Q  P   A  T  F  R   C  V  T   D  R  L
CCATCCAGTT TGGCAACTAC AAAAAGTAGA GGCAGCTGCA GCCACCGCGG  1300
A  I  Q  F   G  N  Y   K  K (Stop codon)
GGCCTCAGTG GGGGTCTCTG GGGAACACAG CCAGACATCC TGATGGGGAG
GCAGAGCCAG GAAGCTAAGC CAGGGCCCAG CTGCGTCCAA CCCAGCCCCC  1400
CACCTCGGGT CCCTGACCCC AGCTCGATGC CCCATCACCG CCTCTCCCTG
GCTCCCAAGG GTGCAGGTGG GCGCAAGGCC CGGCCCCCAT CACATGTTCC  1500
CTTGGCCTCA GAGCTGCCCC TGCTCTCCCA CCACAGCCAC CCAGAGGCAC
CCCATGAAGC TTTTTTCTCG TTCACTCCCA AACCCAAGTG TCCAAGGCTC  1600
CAGTCCTAGG AGAACAGTCC CTGGGTCAGC AGCCAGGAGG CGGTCCATAA
GAATGGGGAC AGTGGGCTCT GCCAGGGCTG CCGCACCTGT CCAGACACAC  1700
ATGTTCTGTT CCTCCTCCTC ATGCATTTCC AGCCTTTCAA CCCTCCCCGA
CTCTGCGGCT CCCCTCAGCC CCCTTGCAAG TTCATGGCCT GTCCCTCCCA  1800
GACCCCTGCT CCACTGGCCC TTCGACCAGT CCTCCCTTCT GTTCTCTCTT
TCCCCGTCCT TCCTCTCTCT CTCTCTCTCT CTCTCTCTCT CTTTCTGTGT  1900
GTGTGTGTGT GTGTGTGTGT GTGTGTGTGT GTGTGTGTGT CTTGTGCTTC
CTCAGACCTT TCTCGCTTCT GAGCTTGGTG GCCTGTTCCC TCCATCTCTC  2000
CGAACCTGGC TTCGCCTGTC CCTTTCACTC CACACCCTCT GGCCTTCTGC
CTTGAGCTGG GACTGCTTTC TGTCTGTCCG GCCTGCACCC AGCCCCTGCC  2100
CACAAAACCC CAGGGACAGC GGTCTCCCCA GCCTGCCCTG CTCAGGCCTT
GCCCCCAAAC CTGTACTGTC CCGGAGGAGG TTGGGAGGTG GAGGCCCAGC  2200
ATCCCGCGCA GATGACACCA TCAACCGCCA GAGTCCCAGA CACCGGTTTT
CCTAGAAGCC CCTCACCCCC ACTGGCCCAC TGGTGGCTAG GTCTCCCCTT  2300
ATCCTTCTGG TCCAGCGCAA GGAGGGGCTG CTTCTGAGGT CGGTGGCTGT
CTTTCCATTA AAGAAACACC GTGCAACGTG AAAAAAAAAA AAAAAAAAAA  2400
A     (PolyA signal)                (PolyA site)
```

Fig. 6.3. *Homo sapiens* biglycan mRNA (Strausberg, R. L., et al. 2002. *Proc. Natl. Acad. Sci. USA* 99, 16899-16903; GenBank BC002416).

(3) Translation starts at the ATG start codon at position 173 and stops at the TAG stop codon at position 1277. The protein obtained from translation starts with Met1 and ends with Lys368. This translated protein is a prepro-protein that undergoes further posttranslational processing.

(4) The N-terminus of the protein contains a signal peptide (translated from the signal sequence in the mRNA) of 16 amino acids that enables the protein to be secreted (transported across a cell membrane). The signal peptide is removed as the protein is secreted.

(5) Immediately following the signal peptide is a 21-amino acid sequence. This short sequence plays a role in folding the polypeptide chain into the correct structure. This stretch of amino acids is cleaved during secretion. This sequence is referred to as the pro-sequence in

this case, because the proceeding signal peptide is called the pre-sequence.

(6) The resulting protein, after the cleavage of the prepro-sequence, is the "mature" protein (the functional protein) which has Asp as the N-terminal amino acid and Lys at the C-terminus with a total of 331 amino acids.

(7) If we perform a computer analysis on the nucleotide sequence for *Hin*dIII restriction sites (A^AGCTT), we will find that the enzyme cuts at positions 291 and 1557, giving 291, 844, and 1266 bp fragments. Agarose gel electrophoresis of the digested DNA will show 3 bands corresponding to these fragment sizes (see Section 7.1).

Does one need to know all these details of the genomic and cDNA sequences of a gene for the purpose of cloning? In fact, one needs only the structural gene sequence (from the Met codon to the stop codon). The transcriptional and translational elements upstream of and downstream of a gene sequence are often not used in cloning. A wide selection of vectors has been developed for cloning. These vectors are constructed with unique promoter, signal sequence, multiple cloning sites, and other control elements for various gene cloning and expression strategies (see Chapter 9). However, knowledge on the organization of a gene enables one to understand the why and how, and to optimize the approach for successful cloning.

Review

1. Referring to the *E. coli dut* gene sequence (Fig. 6.1), list all the transcriptional and translational elements separately, and describe their primary functions.
2. Repeat the same using the biglycan gene sequences (Fig. 6.2 and 6.3).
3. In cloning the biglycan gene for expression, which host system would be the most appropriate to use? *E. coli* or yeast?
4. What is a "Kozak" sequence? What is its function? Can you locate the sequence in (A) the *E. coli dut* gene, and (B) the biglycan gene?
5. Compare the sequences in Figs. 6.2 and 6.3. Highlight the segments in the genomic sequence that match with the mRNA (cDNA) sequence.
6. Do all the exons in the genomic sequence appear in the mRNA? Why or why not? Explain you answer.

Part Two

Techniques and Strategies of Gene Cloning

CHAPTER 7

Enzymes Used in Cloning

The manipulation of DNA utilizes a number of enzymes. These enzymes are naturally occurring in cells involved in transcription, translation, replication and other biological processes. The reactions catalyzed by these enzymes have become an essential part of gene cloning. Examples of enzyme uses in cloning include cutting and joining DNA, deletion or extension of DNA, generating new DNA fragments, and copying DNA from RNA. These enzymes are available commercially in highly purified forms suitable for cloning work.

7.1 Restriction Enzymes

Restriction enzymes are endonucleases that cut internal phosphodiester bonds at specific recognition sequence. Recognition sequences are known as restriction sites in a DNA molecule. There are about 100 restriction enzymes, each recognizing a specific sequence of 4, 5, 6, or 7 nucleotides.

Restriction enzymes are also distinguished according to their mode of action. Some enzymes, such as *Hae*III, cut both DNA strands at the same position, resulting in blunt-end DNA fragments. Many enzymes, however, cut DNA strands at the restriction site at different positions, resulting in DNA fragments with cohesive (sticky) ends. *Hin*dIII, *Sau*I and *Pst*I belong to this class of restriction enzymes. *Hin*dIII cut yields a 5' cohesive end with the 5' end protruding out, whereas *Pst*I cut yields a 3' cohesive end with the 3' end protruding out (Fig. 7.1).

It is not uncommon that the same restriction site occurs more than once in a DNA molecule. Thus digestion by a restriction enzyme often results in a number of DNA fragments. Once the sequence of a DNA molecule is known, it is useful to generate a complete restriction map revealing all the possible restriction sites for a set of common restriction enzymes. This can be conveniently done using computer software. Knowledge of a restriction map facilitates the correct selection of restriction enzymes for DNA manipulation and recombinant DNA construction.

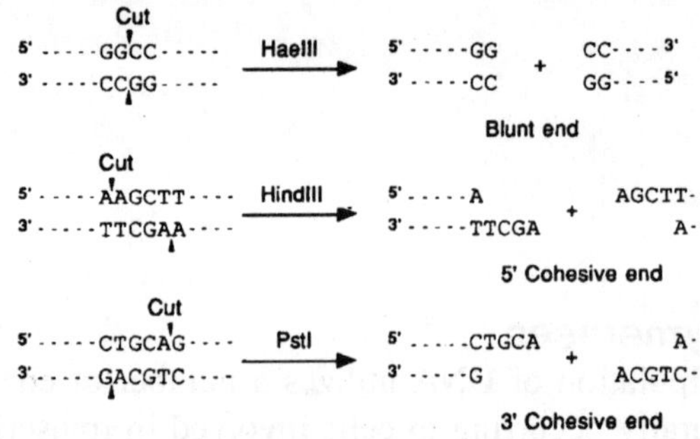

Fig. 7.1. Cutting DNA using restriction enzymes.

7.2 Ligase

Two DNA fragments with complementary cohesive ends can base pair to piece the ends together. The gap between the 3'-OH and the 5'-P is a nick that can be completed by the formation of a phosphodiester bond using bacteriophage T4 DNA ligase in the presence of ATP (Fig. 7.2). The enzyme also acts on blunt end ligation, but with lower efficiency for the lack of base pairing in the ends.

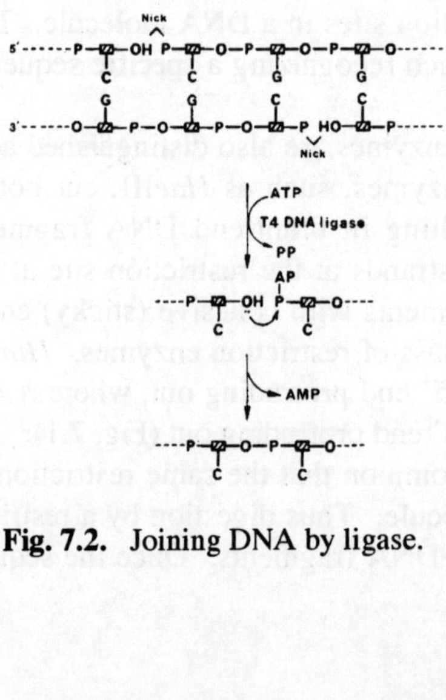

Fig. 7.2. Joining DNA by ligase.

The combined use of restriction enzymes and DNA ligase enables the cutting of DNA at desirable locations and the rejoining of any two or more DNA fragments together. Specific restriction sites can be created by ligating to the DNA molecule a short DNA segment with preformed cohesive ends or containing a specific restriction site. The former is called an adaptor, and the latter is a linker. A blunt-ended DNA ligated with linkers or adaptors can generate new cohesive ends complementary with the ends of another DNA fragment. A caution in using linkers or adaptors is that the product carries an addition of nucleotides, which need to be taken into consideration in the construction of recombinant DNA.

7.3 DNA Polymerases

DNA polymerases are a group of enzymes used very often in gene cloning. It includes DNA-dependent DNA polymerases (*E. coli* DNA polymerase I, T4 and T7 DNA polymerase, *Taq* DNA polymerase) and RNA-dependent polymerases (reverse transcriptase).

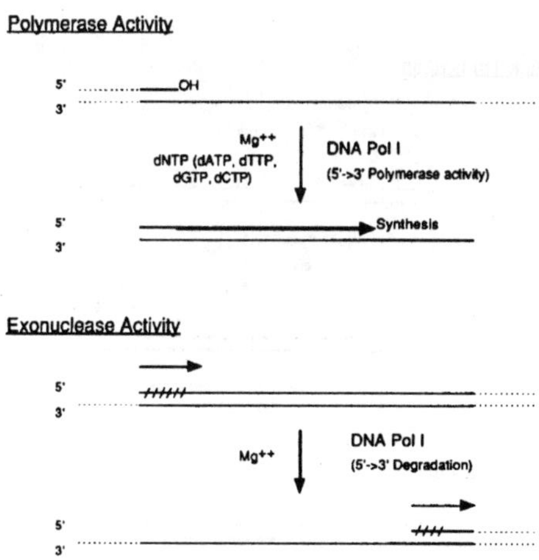

Fig. 7.3. Polymerase activity and exonuclease activity of *E. coli* polymerase I.

7.3.1 *E. coli* DNA Polymerase I

E. coli DNA polymerase I (Pol I) catalyzes the addition of nucleotides to the 3' end of a DNA primer that is hybridized to a ssDNA template. In practice, the short strand of the dsDNA acts as a DNA primer for the complementary strand, which is the template (Fig. 7.3). In addition to the 5'->3" polymerase activity just described, the enzyme also contains a 5'->3' exonuclease activity and a 3'->5' exonuclease activity.

The polymerase activity of DNA polymerase I is utilized in the polymerase chain reaction (PCR) for *in vitro* selective amplification of specific regions of a DNA molecule. The enzyme used in this case is Taq DNA polymerase I isolated from *Thermus acquaticus*, which has high polymerase activity, contains no 3'->5' exonuclease activity, and is more resistant to thermal denaturation than the *E. coli* enzyme (see Section 8.9).

The Klenow Fragment. The enzyme Pol I can be cleaved to produce a large fragment known as the Klenow fragment which contains only the DNA polymerase activity and low 3'->5' exonuclease activity, but lacks the 5'->3' exonuclease activity. The Klenow fragment is the enzyme commonly used for labeling the 3' end of DNA with radioactive nucleotides. It is also the enzyme of choice for nick translation to produce uniformly radioactive DNA. Radiolabeled DNA probes are used to "probe" regions of the same sequence in a recombinant DNA molecule by hybridization (see Section 8.6).

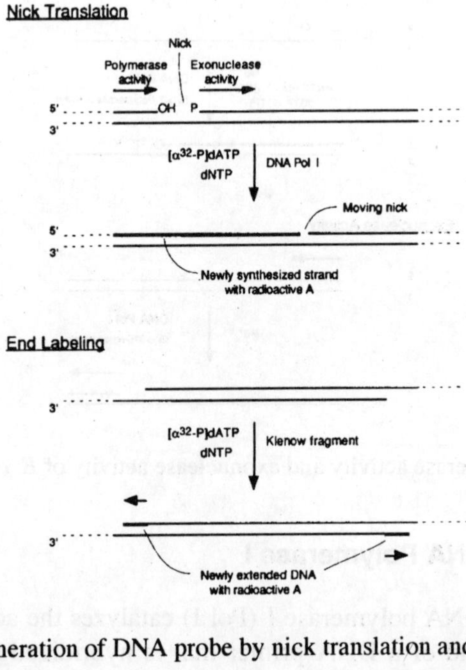

Fig. 7.4. Generation of DNA probe by nick translation and end labeling.

In nick translation, the DNA fragment to be labeled is first nicked in a random manner by the action of pancreatic deoxyribonuclease I (DNase I). At the nick, the two DNA polymerase activities (5'-3' polymerase activity and 5'-3' exonuclease activity) couple to incorporate nucleotides to the exposed 3' end of the nick, while the 3'->5'exonuclease activity of the enzyme degrades the 5' end of the nick. The term "translation" refers to the movement of the nick along the DNA molecule, as polymerization and degradation proceeds. If any of the nucleotides (dNTP) incorporated in the reaction is labeled with ^{32}P (i.e. $[\alpha$-$^{32}P]$dNTP), the DNA molecule carrying the labeled nucleotide becomes radioactive (Fig. 7.4). In practice, only one of the four dNTP (dATP, dTTP, dGTP, and dCTP) is labeled, for example, $[\alpha^{32}P]$dATP. Non radioactively labeled nucleotides, for example, fluorescein-, rhodamine- and coumarin-dUTP, can be used instead. In this case, the resulting probe contains fluorescent tags that can be detected. In nick translation, the existing nucleotide sequence is renewed without net synthesis occurring, resulting in a complete replacement of the sequence uniformly labeled.

7.3.2 Bacteriophage T4 and T7 polymerase.

E. coli, infected by bacteriophage T4 or T7, produces DNA polymerases, known as bacteriophage T4 or T7 DNA polymerase. T4 polymerase possesses a very active single-stranded 3'->5' exonuclease (activity many folds stronger than that of the Klenow fragment), but lacks a 5'->3' exonuclease activity. This enzyme is frequently used to fill 5' protruding ends with labeled or unlabeled dNTPs or to generate blunt ends from DNA with 3' overhangs (Fig. 7.5).

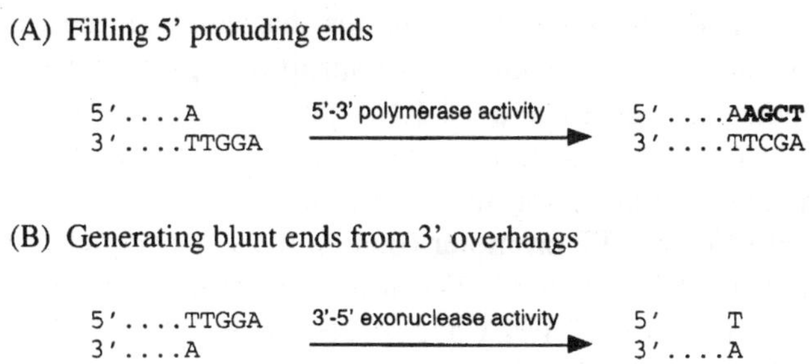

Fig. 7.5. Filling a 5' protruding end (A) and converting a 3' overhang to a blunt end (B) using T4 DNA polymerase.

The T7 polymerase native enzyme has very high 3'->5' exonuclease activity in addition to its polymerase activity. The enzyme used today is chemically or genetically modified to have low 3'->5' exonuclease activity, high processivity, and fast polymerase rate. These properties make it suitable for use in DNA seequencing (see Section 8.8).

7.3.3 Reverse Transcriptase

Reverse transcriptase is a RNA-dependent DNA polymerase. The enzyme uses RNA as a template to synthesize a complementary DNA strand to yield a RNA:DNA hybrid. The enzyme reaction requires a DNA or RNA primer with a 3'-OH (Fig. 7.6). Reverse transcriptase is used for the synthesis of the first strand cDNA in the construction of cDNA libraries for gene isolation (see Section 11.2).

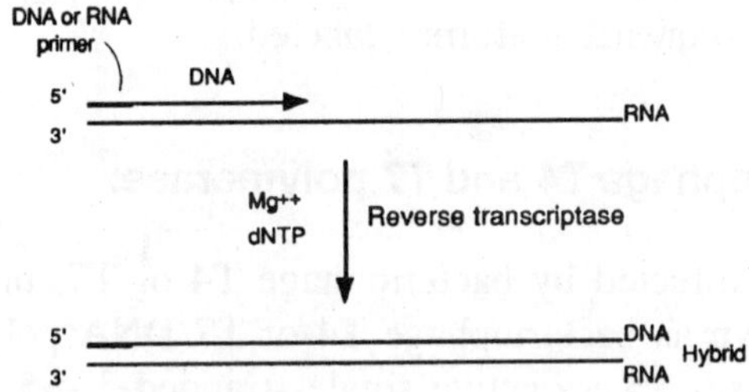

Fig. 7.6. Synthesis of complementary DNA using a RNA template.

7.3.4 RNA Polymerases

These are DNA-dependent enzymes that catalyze the transcription of DNA sequences downstream of an appropriate promoter (see Section 5.2). The bacteriophage T7 and T3 RNA polymerases are used in *in vivo* expression of cloned genes; transcription is very rapid with high possessivity. In the expression vector, the gene of interest, for example, is cloned under the control of the T7 promoter (that is, the gene sequence is inserted downstream of the T7 promoter). When the construct is transformed into a suitable *E. coli* host, the T7 RNA polymerase poduced by the host induces the expression of the gene (see Section 9.1.1).

7.5 Phosphatase and Kinase

Ligation of two DNA fragments requires DNA molecules containing 5'-P and 3'-OH termini. Removal (dephosphorylation) or re-addition (phosphorylation) of the phosphate residue by enzyme action is often used to regulate ligation reaction (Fig. 7.7). Alkaline phosphatase (from E. *coli* or calf intestine) removes phosphate residues from the 5' terminus. Bacteriophage T4 polynucleotide kinase catalyzes the transfer of the γ-phosphate of ATP molecule to the 5' terminus of a DNA fragment. Kinase is also used for radiolabeling DNA, particularly short DNA fragments. In this case, the enzyme transfers the radioactive γ-phosphate of [γ-^{32}P]dNTP to the 5' end of the DNA fragment.

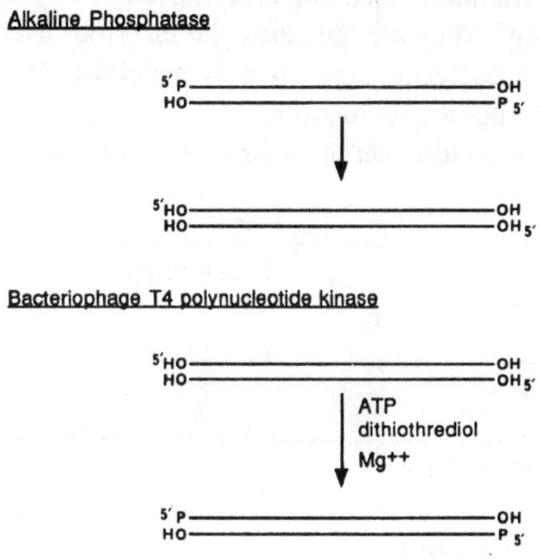

Fig. 7.7. Dephosphorylation and phosphorylation of DNA.

Review

1. Given the following DNA fragment, design an adapter to yield a 5' cohesive end.
```
5'--GCTG--3'
3'--CGAC--5'
```
2. Referring to the DNA fragment in problem 2, how do you design a linker to add a *Hin*dIII restriction site to the DNA fragment?
3. How do you ligate the following two DNA fragment using cohesive ends?
```
5'--GCTA        GCAG--3'
3'--CGAT        CGTC--5'
```

4. Write the restriction fragments, and indicate whether the fragments are blunt ends, 3' cohesive ends, or 5' cohesive ends.

*Xba*I	5'--TCTAGA--3' 3'--AGATCT--5'
*Sma*I	5'--CCCGGG--3' 3'--GGGCCC--5'
*Apa*I	5'--GGGCCC--3' 3'--CCCGGG--5'
*Msp*I	5'--CCGG--3' 3'--GGCC--5'

5. What are the two methods used for labeling DNA? What are the enzymes employed in labeling? Why are the particular enzymes used?
6. Is it possible to use bacteriophage T4 polynucleotide kinase for end-labeling DNA fragments? Explain your answer.
7. Label the level of activities (high or low or none) for the following polymerase enzymes.

Activity	Polymerase	Exonuclease 5'->3"	Exonuclease 3'->5'
E. coli polymerase I			
Bacteriophage T4 polymerase			
Bacteriophage T7 polymerase			
Taq polymerase			

CHAPTER 8

Techniques Used in Cloning

The theoretical and experimental background for the cloning techniques is closely tied with the biological processes described in Part 1. Volumes of protocols are available for use in gene cloning. Fortunately, the fundamentals of the techniques involved are not difficult to understand.

8.1 DNA Isolation

The most commonly used technique in cloning is to isolate and purify plasmid DNA from transformed *E. coli* cultures. This procedure, known as miniprep, requires as little as 1 ml overnight culture, and is performed by alkaline lysis of the cultured cells. The alkaline solution of NaOH/SDS breaks up the cell walls, releasing the cellular contents into solution. This is followed by neutralization with potassium acetate, and then precipitation of the DNA by 95% ethanol. In recent years, this procedure has been standardized and simplified by the use of mini-prep columns employing membrane/resin technology. A wide selection of kits is now available providing a fast, simple, and cost-effective way for the isolation of plasmid DNA, genomic DNA, and RNA from microbial, fungal, plant, and mammalian cells.

8.2 Gel Electrophoresis

Gel electrophoresis is an important technique for the separation of macromolecules. DNA molecules of different sizes can be resolved by agarose gel electrophoresis, whereas proteins are usually separated by polyacrylamide gel electrophoresis (PAGE).

8.2.1 Agarose Gel Electrophoresis

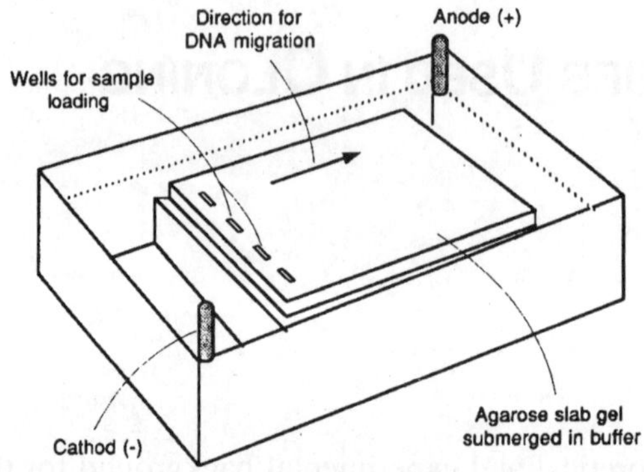

Fig. 8.1. Agarose gel electrophoresis apparatus for separation of DNA fragments.

For DNA separation, the gel matrix of choice is agarose, although acrylamide is sometimes used, particularly for large DNA fragments. The DNA sample (e.g a restriction digest of a mini-prep) is loaded in wells at one end of the agarose gel slab which is submerged under buffer. When an electric field is applied, the DNA fragments in the digest migrate to the anode (+) electrode. The smaller the DNA fragment is, the faster it migrates through the agarose gel due to the effect of molecular sieving (Fig. 8.1). Hence, DNA fragments differing in lengths (sizes) can be separated into discrete bands.

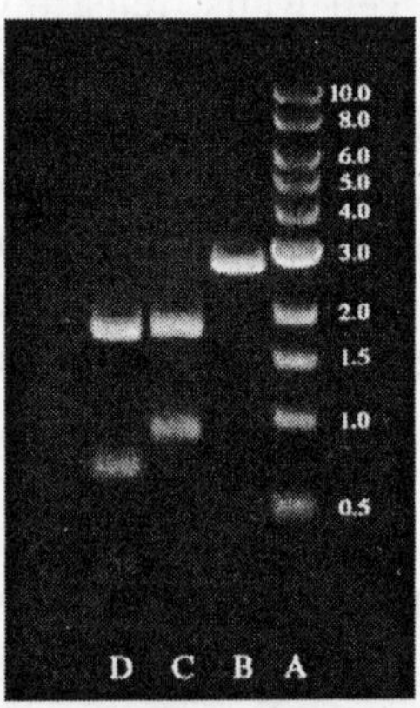

Fig. 8.2. Resolved DNA bands in agarose gel after electrophoresis and ethidium bromide staining. Gel viewed under ultraviolet light. Lane A: DNA marker, Lanes B, C, D: pUC19 plasmid cut by *Bam*HI, *Pvu*I, and *Rsa*I respectively.

The DNA bands are visualized by staining with ethidium bromide, and viewed under ultraviolet light. It is common practice to do a parallel run with a size marker (a mixture of several DNA fragments of known sizes). The size of an individual DNA band can be estimated by comparing the distance of migration with that of the known fragments in the marker (Fig. 8.2).

8.2.2 Polyacrylamide Gel Electrophoresis

Proteins can similarly be separated by gel electrophoresis. The gel matrix used is acrylamide, and the staining reagent is Comassie blue. For high sensitivity, silver staining can be employed. Protein samples are run parallel with a protein marker consisting of a number of standard proteins with known molecular sizes. The resolved protein bands are analyzed for their molecular sizes by comparing the distance of migration with that of the known fragments in the marker (Fig. 8.3).

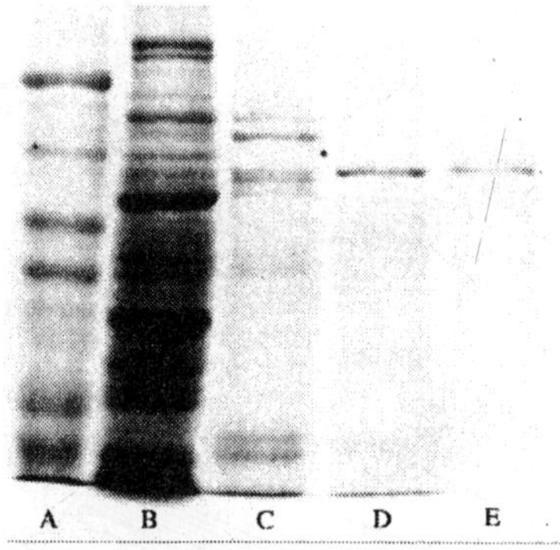

Fig. 8.3. Resolved protein bands in polyacrylamide gel after electrophoresis and coomassie blue staining. LaneA; molecular-weight marker containing proteins of known sizes. Lane B: Crude protein extract from tissue. Lane C, D, E: Progressive purification of the protein of interest.

PAGE is particularly useful for analyzing gene expression. Recombinant proteins produced by a clone may be identified by comparing electrophoretic bands of the cell extract of the clone versus that of the control (transformant without the gene insert). However, cell extracts usually contain many proteins that may simply create a smear instead of appear as discrete bands on the gel, if protein stains are used. It is often desirable to at least partially purify the protein or better yet to conduct a Western blot for immunodetection (see Sections 8.3 and 8.7).

8.3. Western Blot

To confirm the expression of a recombinant protein, the first step is to separate the proteins in an extract of the host cell tissue by PAGE. The resolved protein bands in the gel are transferred to a nitrocellulose membrane by a technique called Western blot. The protein gel and membrane are sandwiched between filter paper, submerged in a tank buffer, and subjected to an electric current. Bands migrating out of the gel are bound onto the adjacent membrane. The protein bands on the membrane are subjected to immunological detection (see Section 8.7). Since the antibody is antigen-specific, it will bind only to the expressed recombinant protein among the many proteins in the cell extract, and be visualized as a single band. The band that reacts positively with the antibody should also match with the molecular size predicted for that protein.

8.4 Southern Transfer

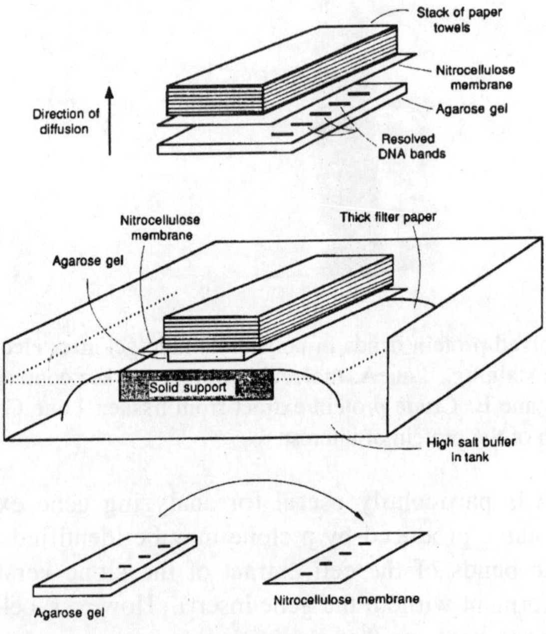

Fig. 8.4. Transfer of DNA bands from an agarose gel to a nitrocellulose membrane.

In a similar technique for DNA, resolved DNA bands in an agarose gel after electrophoresis can be transferred to a membrane (generally nitrocellulose or nylon type) by a technique known as Southern transfer (or blot) (Fig. 8.4). The process in this case is a simple diffusion assisted by pulling a high salt buffer through the gel and nitrocellulose membrane. Alternatively, an electric field can be applied to increase the transfer rate with similar results. A particular DNA fragment (a band) on the membrane can be identified by hybridization (see Section 8.6)

8.5 Colony Blot

Colony blot is a variation of Southern blot. Instead of working with resolved DNA bands, one simply uses a nitrocellulose membrane to directly transfer bacterial colonies (also works for bacteriophages), obtaining a replicate of the colonies grown on a petri dish. The colonies attached to the membrane are subjected to alkali hydrolysis and detergent treatment to release the DNA content from the bacterial cells, which would then bind to the membrane (Fig. 8.5).

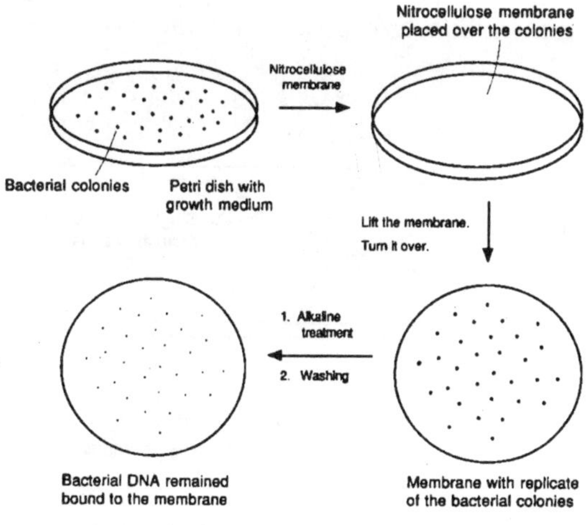

Fig. 8.5. Transfer of DNA from bacterial colonies to a nitrocellulose membrane.

It is also common to selectively pick the colonies and arrange them in a fresh petri dish. A nitrocellulose membrane is then laid onto the plate, followed by a short incubation. The membrane is then lifted and alkali-treated for cell lysis (Fig. 8.6).

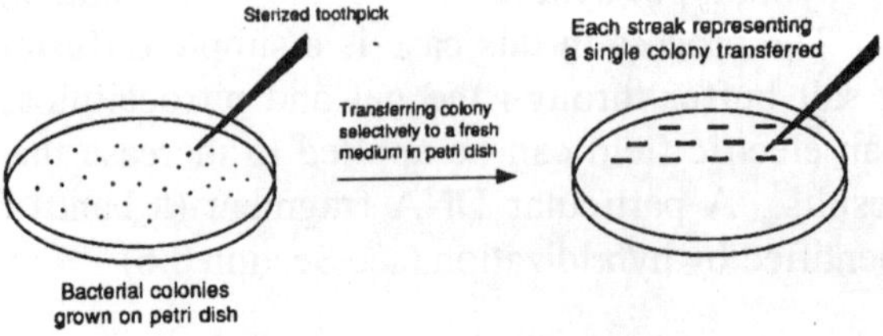

Fig. 8.6. A variation of colony blot.

8.6 Hybridization

The primary purpose of conducting a blot, in most cases, is for DNA-DNA hybridization, with labeled DNA probe. Hybridization is a DNA denaturation-renaturation process. The DNA on the nitrocellulose membrane is first denatured, and the ssDNA then binds to the radiolabeled probe (also denatured) if they are complementary. The DNA probe is labeled usually with ^{32}P by nick translation or end labeling (see Sections 7.3.1), although non-radioactive probes are increasing being used (see Section 8.11).

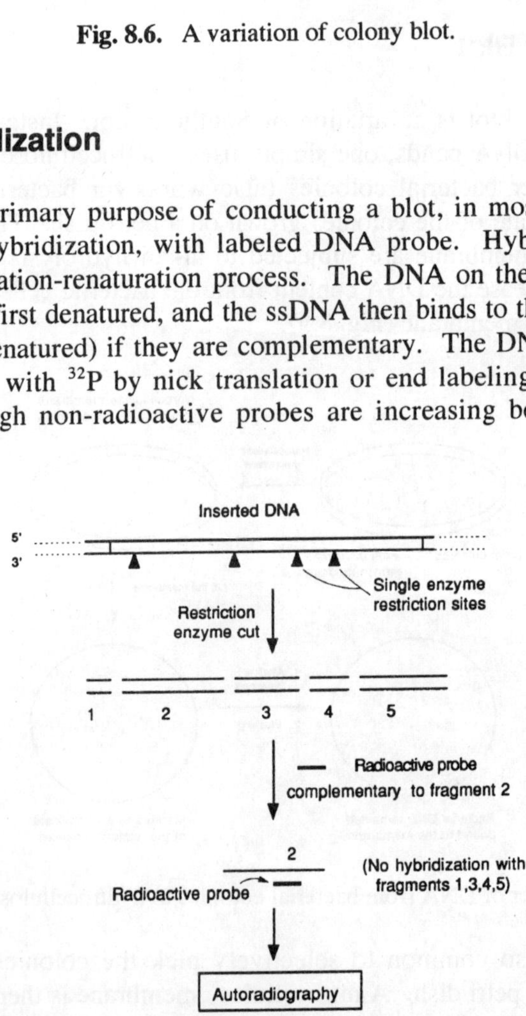

Fig. 8.7. Identifying specific DNA fragments by hybridization.

After hybridization, the membrane is exposed to X-ray film, followed by film development. Only bands that are hybridized with the probe appear on the film. The comparative band position, and therefore the identity of the band can be traced in the original gel (Figs. 8.7, 8.8).

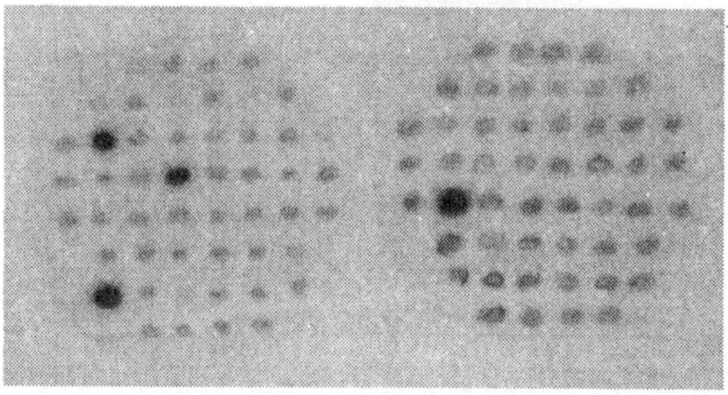

Fig. 8.8. An autoradiogram after colony blot and hybridization. Colonies containing DNA fragments that hybridized with the DNA probe showed highly intensified signals compared with the light background.

The procedure of hybridization implies that at least, a short portion (~18 bp) of the sequence must be known, so that a synthetic oligonucleotide can be made for the probe. Amino acid sequence of the gene product, if known, can be used to deduce the DNA sequence for synthesis of oligonucleotides.

Hybridization is one of the techniques to enable the identification and isolation of a clone containing the DNA of interest in a vast population of clones in a transformation. Hundreds of clones can be screened in a high-throughput manner for the target clone in a single experiment. This is a conventional technique used for isolating clones containing the gene of interest from cDNA or genomic libraries (see Sections 11.1 and 11.2). In many cases, PCR has become a common alternative to hybridization (See Section 8.9).

8.7 Immunological Techniques

Frequently after cloning the gene of interest, one would like to check for expression. A common method is to detect the gene product using immunological techniques. This method requires that: (1) the gene must be properly constructed and inserted for expression because it is the protein targeted for detection, and (2) the protein needs to be isolated and purified from its natural source, which is then used to raise antibodies. Alternatively, antibodies can be raised against a peptide epitope designed from the primary sequence of the protein. This step of raising antibodies can be eliminated, if the gene is constructed in fusion with a His-tag in the vector, as anti-His antibodies which are available commercially, can then be used (see Section 9.1.1).

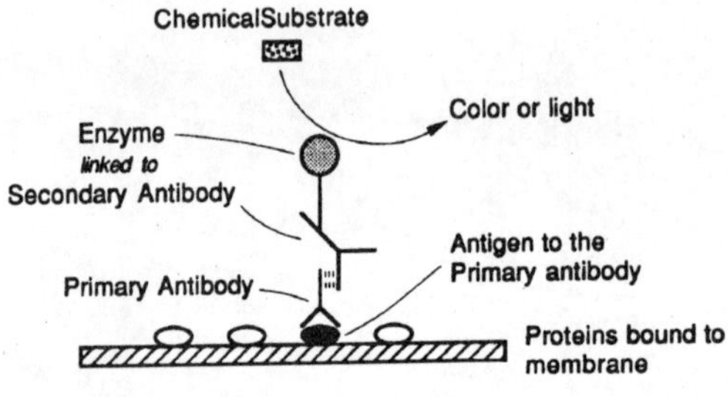

Fig. 8.9. Identifying specific proteins by immunological techniques.

The antibodies generated bind to the target protein on the membrane (from Western blot) via the formation of an antibody-antigen complex. In the case of anti-His antibodies, the antibody binds to the His-tag, which is part of the fusion protein. The target protein is called the antigen, and the antibody that binds to the specific antigen is known as the primary antibody. A secondary antibody is then used to bind to the first antibody. This secondary antibody is tagged with an enzyme that would initiate a chemical signal, in the presence of an appropriate substrate, leading to color development. This technique is known as ELISA (enzyme-linked immunosorbent assay) (Fig. 8.9 and 8.10). For high sensitivity, chemiluminescent substrates can be used (see Section 8.11). The secondary antibody may also be labeled, for example with ^{125}I, and detection is achieved by autoradiography.

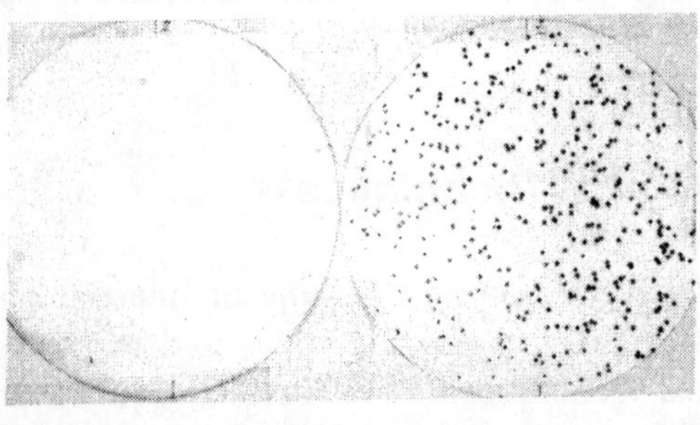

Fig. 8.10. Immunological screening of λDNA library. Left membrane: A putative clone was identified. Right membrane: The clone was propagated for a secondary screening to confirm the result.

8.8 DNA Sequencing

After a gene is isolated, a complete DNA sequence of the DNA molecule is needed if little information exists about the gene. In cases where the amino acid sequence of the gene product is known, the end segments of the isolated gene should still be analyzed. After all, it is necessary to confirm that the isolated gene has the expected sequence. Moreover, the 5' end portion must be sequenced for proper construction and insertion of the gene into a suitable cloning vector.

For quite a while, DNA was sequenced as ssDNA, such as obtained by M13 cloning (see Section 9.1.2). Nowadays, DNA is sequenced as denatured dsDNA. Sanger's dideoxy chain termination method is the most widely adopted sequencing procedure. The method begins with the synthesis of a complementary strand of the ssDNA using the enzyme, *E. coli* polymerase I Klenow fragment (or more frequently modified T7 DNA polymerase) and a mixture of dNTP (dATP, dTTP, dGTP, dCTP). A DNA primer is also needed for the enzyme to act. In the synthesis process, the nucleotides (dNTP) are added in a sequence complementary to the DNA being sequenced. However, if 2',3'-dideoxyribonucleoside triphosphate (ddNTP = ddATP, ddTTP, ddGTP and ddCTP) which lacks the free OH group at the carbon 3' position is present in the reaction mixture, polymerization will terminate if a ddNTP is picked up any time during the synthesis (Fig. 8.11).

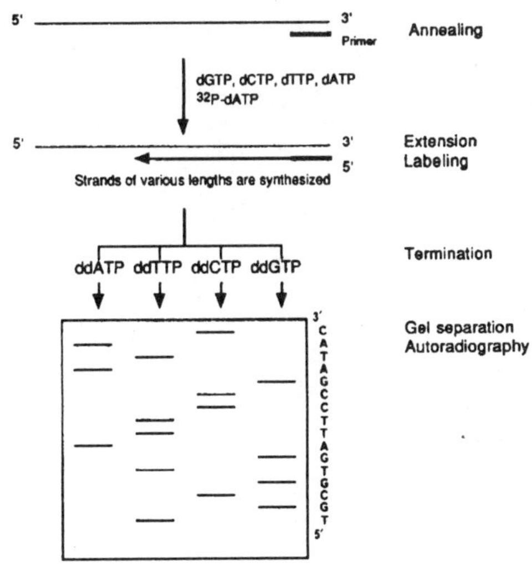

Fig. 8.11. DNA sequencing by the Sanger dideoxy method.

In DNA synthesis, polymerization of the DNA strands occurs at different speed. Consequently, a heterogeneous mixture of DNA strands of various lengths is produced that is representative of all possible sequences. Depending on which ddNTP is used, the termination nucleotide will be either A, or T, or G, or C. In fact, all four reactions are carried out separately, generating 4 groups of sequences, one group with all endings in ddATP, one group with all endings in ddTTP, one group with endings in ddGTP, and one group with endings in ddCTP.

DNA segments in each group can be separated by polyacrylamide gel electrophoresis. However, the resolved bands also need to be visualized or detected. A simple way to do this is to incorporate radiolabled nucleotides, such as [α-^{33}P]dATP, into the reaction mixture, so that all the fragments are radiolabeled. Now the resolved bands can be detected by X-ray autoradiography (Fig. 8.12).

The bands are read from bottom up in a 5' to 3' direction. Usually the bands at the bottom are too faint and the bands at the top are too compressed to read. There are only a limited number of bands (~200 bases) that can be read with certainty. In order to determine a complete sequence of a gene, the procedure has to be repeated a number of times moving the primer along the sequence.

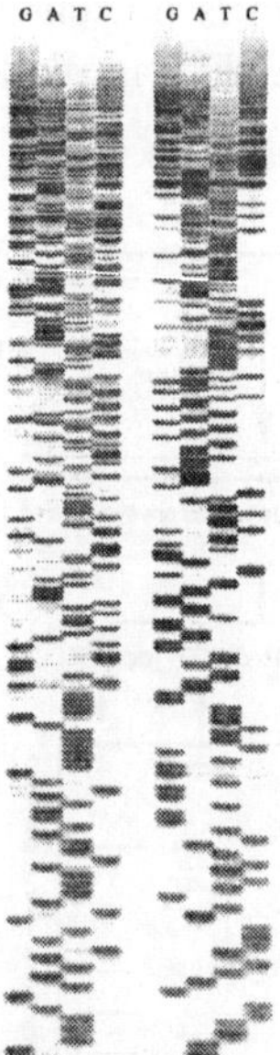

Fig. 8.12. Resolved bands in sequencing gel after electrophoresis and autoradiography.

In the last decade, manual sequencing described above has largely been replaced by automated sequencing. A major change in the protocol is the use of flourescent labels instead of isotopes. Four flourescent dyes are used, with one for each ddNTP. Chains terminated with A are labeled with one flourescent dye, chains terminated with T are labeled with a second dye, chains terminated with G is labeled with a third dye, and chains terminated with C is labeled with a fourth dye. Using 4 different flourescent labels, it is now possible to carrying out the four sequencing reactions as in the manual protocol in a single tube, and to load the one reaction mixture into one lane of the polyacrylamide gel. The flourescence detector scans the separated bands, discriminates between the four different flouresent labels, and translate the results into a readable chromatogram of peaks distinguished by four colors: green for A, red for T, black for G, and blue for C (Fig. 8.13). Automated sequencing in ordinary lab research can read up to about 600 bp in a single run.

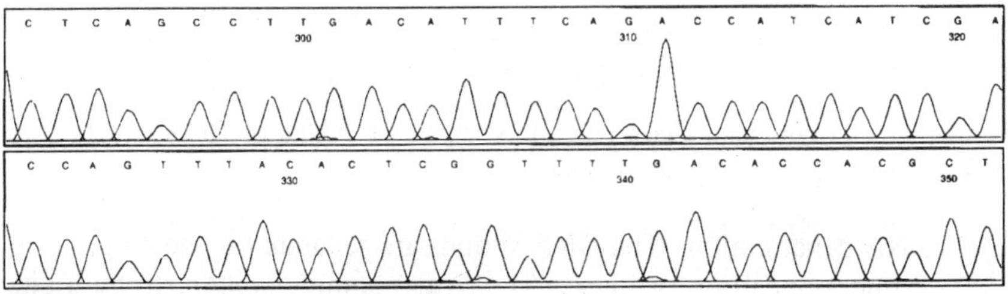

Fig. 8.13. Chromatogram of autosequencing. In the original chromatogram, peak lines are distinguished by different colors: green for A, red for T, black for G, and blue for C.

8.9 Polymerase Chain Reaction

The polymerase chain reaction (PCR) utilizes the polymerase activity of the enzyme DNA polymerase I to amplify a chosen segment of a DNA molecule. Any segment in a DNA molecule can be chosen, as long as the short sequences in the 5' and 3' flanking regions of the segment are known. This information is needed for synthesizing short oligonucleotides used as DNA primers in the polymerase reaction. Notice that the primers need not to be phosphorylated. The design of primers for optimizing PCR reactions can be conveniently performed using computer programs.

PCR consists of cycles of repeating 3 steps: denaturation, annealing and polymerization (Fig. 8.14).

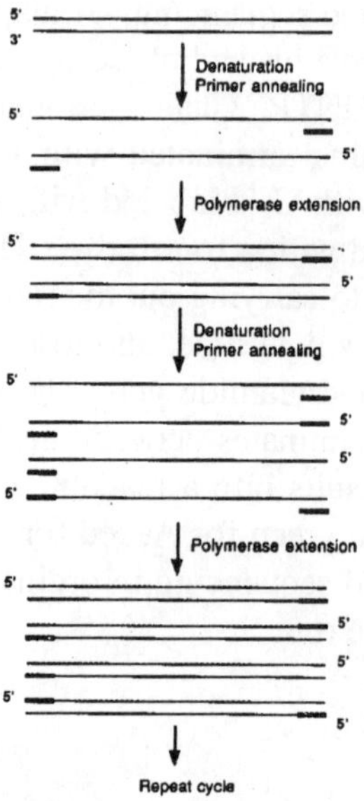

Fig. 8.14. Polymerase chain reaction consists of cycles of DNA denaturation, primer annealing, and polymerase reaction.

(1) DNA denaturation. The DNA strands are separated by heating.

(2) Primer annealing. Primers (~18 bases) complementary to the flanking regions are then annealed to the ssDNA strands by cooling the reaction mixture.

(3) Polymerase Extension. DNA polymerase synthesizes new DNA strands beginning at the primer using the parent ssDNA strands as template.

The cycle is repeated. Each newly synthesized DNA acts as template strands for the succeeding cycle. Consequently, PCR yields an exponential increase of the DNA segment. By the end of the n cycles, the number of copies of DNA = 2^n.

The enzyme used in PCR is Taq polymerase I isolated from a thermphilic bacterium *Thermus acquaticus*. *Taq* polymerase has high polymerase activity, contains no 3'->5' exonuclease activity, and is resistant to denaturation at higher temperatures than the *E. coli* enzyme. The thermostability of the enzyme enables it to be unaffected by repeated heating and cooling cycles. Several thermostable DNA polymerases are now commercially available, some are produced by recombinant techniques.

It is important to note that the polymerase reaction in living cells has a proofreading mechanism to correct errors that occur during base-pairing. This is not present in PCR which is conducted in test tubes. *Taq* polymerase has an error rate of about 10^{-4} (error/bp incorporated). Genetically engineered mutants that exhibit error rates many folds lower than Taq DNA polymerase are now available, if high fidelity is desired. The Pfu polymerase isolated from *Pyrococcus furiosus* appears to have the lowest error rate at roughly 10^{-6}.

PCR may in some cases replaces or supplements the traditional procedures of cloning, culturing, restriction digestion, and purification steps in obtaining a piece of DNA in sufficient quantity for manipulation. The product obtained from PCR is sufficient for identification and quantification by gel electrophoresis. It is now possible to isolate and amplify a gene from a haystack of DNA molecules, with a product often in sufficient amount even for direct sequencing. PCR has found numerous applications in areas such as disease diagnosis (see Chapter 19), DNA typing (see Chapter 20), and environmental and quality control.

8.10 Site-Directed Mutagenesis

The primary objective in protein engineering is to alter (add, substitute, or delete) one or more amino acids of a protein to effect a change in its function. This can be achieved by modifying the corresponding nucleotides of the gene coding for the protein. For example, a change from T to A in the following DNA sequence will result in substitution of the amino acid Phe with Tyr.

```
5′ ---TTA CAA GAC TTT GAA---
N  ---Leu Gln Asp Phe Glu---
```

After site-directed mutagenesis

```
5′ ---TTA CAA GAC TAT GAA---
N  ---Leu Gln Asp Tyr Glu---
```

Mutagenesis is done by hybridization of a synthetic oligobucleotide carrying the mutated nucleotide to the cloned gene, synthesis of the mutant strand by DNA polymerase, and transformation of the DNA hybrid into host cells. Replication carried out by the transformed cells will give DNA molecules with a mutated nucleotide (Fig. 8.15). In some cases, the use of PCR provides a simpler procedure to achieve the same objective, although the basic idea remains in that the mutated nucleotide is carried in the primer (Fig. 8.16).

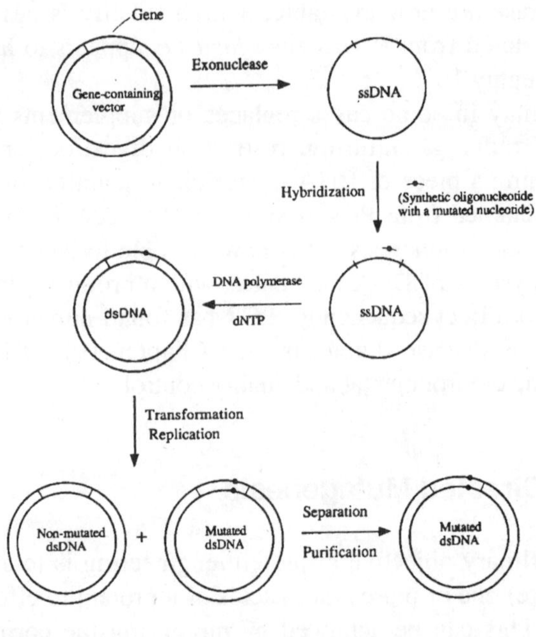

Fig. 8.15. General scheme of site-directed mutagenesis.

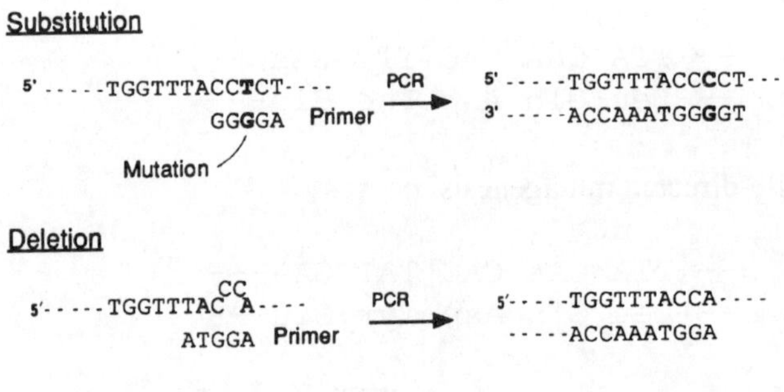

Fig. 8.16. DNA substitution and deletion by PCR.

8.11 Non-radioactive Detection Methods

A common method to label a DNA is by nick translation or end labeling with radioactive nucleotides (see Section 7.3.1). In recent years, non-isotopic labeling has become popular due to increased safety and environmental concerns. Most non-radioactive labeling methods are based on the enzymatic conversion of a chemiluminescent substrate to a stable intermediate compound that decays and emits light (Fig. 8.17). The enzyme

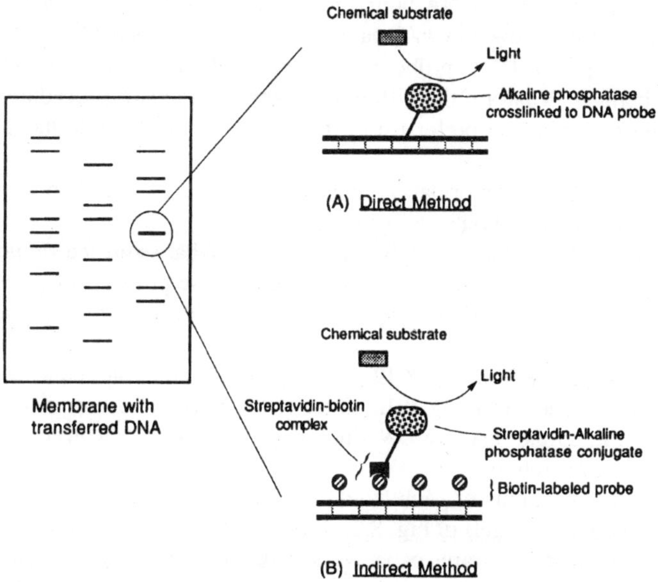

Fig. 8.17. Nonradioactive labeling using chemiluminescence.

used most often is alkaline phosphatase which is either directly or indirectly coupled to the detection system. In the direct method, the enzyme is chemically or enzymatically crosslinked to the DNA probe. In the indirect method, the DNA probe is first conjugated with an organic molecule such as biotin. The resulting biotin-labeled probe is used for hybridization. The probe that is hybridized to the target DNA is then detected by forming a tight complex with a Streptavidin-enzyme conjugate. (Streptavidin is an egg white protein with a very high affinity for biotin.) Chemiluminescence has been used for detection in Southern blot, colony blot, northern blot, and many other applications.

Review

1. What is the unique feature of the structure of DNA or protein that causes the macromolecules to migrate when an electrical field is applied to the gel?
2. Is the concept of hybridization applicable to Western blot? Explain your answer.
3. For hybridization, the DNA bound to the membrane is first denatured. What is denaturation? How do you denature DNA on a membrane?
4. The labeling of a DNA probe can be achieved by either nick translation or end-labeling. What are the differences between the two procedures? Which procedure would you prefer using if the DNA to be labeled has a short sequence?
5. What is the specific information required prior to (A) DNA hybridization, (B) immunological detection?
6. Suppose a M13 vector is used for sequencing. What sequence would you pick for making the DNA primer? Do you generate the primer based on the DNA insert or based on the M13 sequence flanking the insert? Explain your answer.
7. Read the sequence from Fig. 8.12. Specify the 5' end of the sequence.
8. Read the sequence from Fig. 8.13. Specify the 5' end of the sequence.
9. Why is it important that the DNA polymerase used in PCR needs to be stable at to high temperatures?
10. Measure the positions of each band of the DNA marker relative to the origin (use the edge of the gel) in Fig. 8.2. Plot these distances (x-axis) on a piece of three-cycle semilog graph paper versus the known sizes of each fragment. Connect the points to form a curve. Measure the distances of the resolved DNA fragments in the middle three lanes, and use the marker curve just generated to estimate the sizes of the fragments.

CHAPTER 9

Cloning Vectors for Introducing Genes Into Host Cells

The introduction of a foreign DNA into a host cell in many cases requires the use of a vector. Vectors are DNA molecules used to transfer a gene into a host (microbial, plant, animal) cell, and to provide control elements for replication and expression. The vector to be used is determined by the type of host cells and the objectives of the cloning experiment.

9.1 Vectors for Bacterial Cells

9.1.1 Plasmid Vectors

A vast selection of bacterial vectors can be obtained commercially. Bacterial vectors are among the most extensively studied with a wealth of information available for novel manipulation and construction. The most widely used vectors for bacterial cells belong to a group of vectors called plasmid vectors. These vectors have their origin from extra-chromosomal circular DNA, called plasmid, found in certain bacterial cells. Plamid vectors used for cloning are typically less than 5 kb. Large DNA molecules are difficult to handle and often subject to degradation. The efficiency of transformation decreases with increasing size of the plasmid. For cloning use, a plasmid vector preferably contains a number of structural elements.

(1) Replication origin: For replication of the vector DNA utilizing the bacterial host system.

(2) Cloning sites: Plasmid vectors consist of artificially constructed recognition sequences for a number of restriction enzymes. This

cluster of sites, referred to as multiple cloning sites (MCS), serves to facilitate the convenient insertion of a foreign DNA.

(3) Selectable markers: These are usually antibiotic resistance genes, such as ampicillin resistance (Amp^R) and tetracycline resistance (Tet^R). The purpose of having these markers is to screen for transformed cells. Non-transformed cells (not picking up the plasmid vector) will not survive in a growth medium containing the antibiotic.

(4) Some plasmid vectors also contain a promoter region upstream of the multiple cloning site. This construction enables the transcription and translation of the inserted DNA. Of course, care must be taken to ensure the insertion in the proper reading frame (see Section 4.6). Vectors of this type are called expression vectors.

(5) An optional but popular feature often added to a vector is a polyhistidine sequence (e.g. 5'-CACCACCACCACCACCAC encoding 6 histidines) so that the expressed protein will have a short polyhistidine fused either at the N- or C-terminus. The tagging of a protein allows simple one-step purification of the fused protein by nickel columns, because polyhistidine binds to nickel. Other types of tags can also be used, but His-tag is by far the most popular.

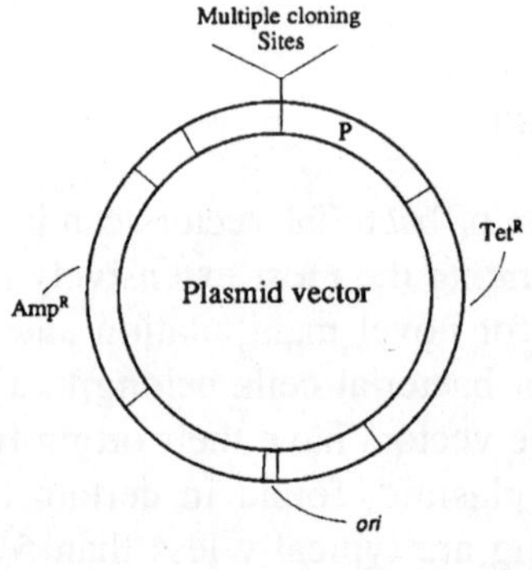

Fig. 9.1. Structural organization of a plasmid vector.

High and Low Copy Plasmids. Plasmids can be grouped into high copy or low copy plasmids, depending on the number of molecules of the plamid found in a bacterial cell. Low copy refers to 1-25 plasmids per cell. High copy refers to 100 copies and more per cell. High copy plasmid vectors are the choice if a high yield of the recombinant DNA is desired. On the other hand, it may not be desirable to use high copy plasmid vectors if the gene product is expected to cause adverse effects to the host cells. Copy number depends on the origin of replication (the DNA region known as the replicon, see Section 4.8), which determines whether the plasmid is under relaxed or stringent control. The copy number is also dependent on the size of the plasmid and its associated insert. The pUC plasmid and its derivatives can reach very high copy numbers within the host cell, while the pBR322 plasmid and those derived from it maintain at very low copy numbers per cell.

pUC Plasmids As An Example. The following description on the pUC series plasmid vectors illustrates how a plasmid vector should operate. The pUC series contain a number of constructions that have become standards in many other vector systems (Fig. 9.2).

(1) The β-lactamase gene (ampicillin resistance, Amp^R), and the origin of replication from the pBR322 plasmid (one of the early *E. coli* cloning vectors that gives rise to many more recently developed vectors) are retained as part of the pUC plasmid vector.

(2) The *lac* operon in pUC contains a truncated *lacZ* (β-galactosidase) gene coding for the N-terminal segment (amino acids 11-41, called α-peptide) of the enzyme. The truncated *lacZ* gene is referred to as *lacZ'*. The *lacI* gene is also truncated, referred to as *lacI'*. Therefore, the *lac* operon in pUC vectors is represented as *lacI'OPZ'* (see Section 5.1).

(3) A cluster of recognition sites for a number of restriction enzymes is inserted into the *lacZ'* region. This cluster of sites is the multiple cloning site (MCS).

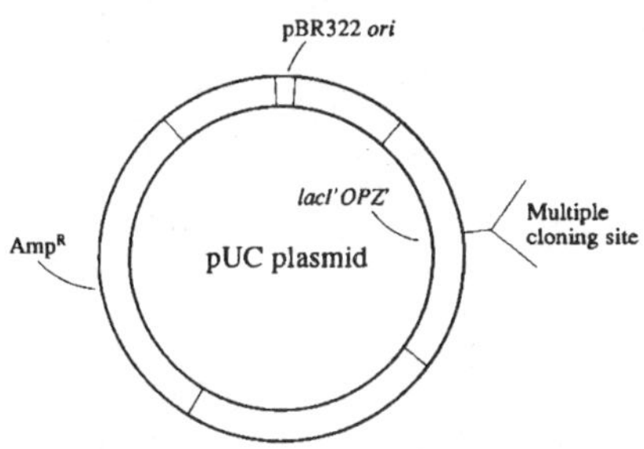

Fig. 9.2. Structural organization of a pUC plasmid vector.

The pUC plasmids are expression vectors, because the *lac* operon is active when isopropyl-β-D-thiogalactopyranoside, IPTG (an analog of lactose, an inducer of *lac* operon) is supplied in the growth medium (see Section 5.2.2). Since the cloning sites are clustered in the *lacZ'* region, the expressed product is a fusion protein carrying a short segment of the β-galactosidase enzyme.

The ampicillin resistance gene is a selectable marker. In a transformation step, only a fraction of the bacterial cells will pick up the vector DNA. The efficiency varies but usually in the range of 0.01%. Following transformation, the cells are plated on a medium containing ampicillin. Majority of the *E. coli* cells are non-transformants, which do not survive for the lack of the ampicillin resistance gene. Only transformants (those that have picked up the plasmid) are ampicillin resistant, and will show up as white colonies (Fig. 9.3).

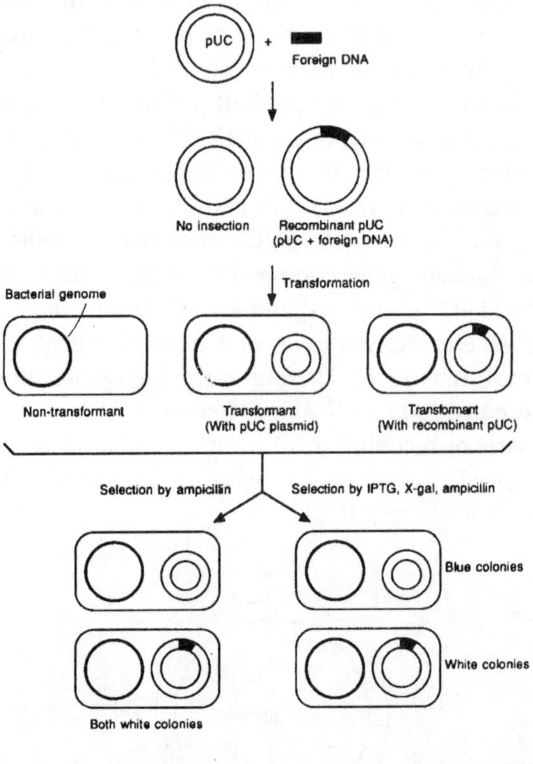

Fig. 9.3. Selection of transformants using a pUC plasmid vector.

However, using the Amp^R gene alone in the selection does not distinguish cells that pick up the pUC vector from those that pick up the recombinant vector DNA (i.e. pUC vector containing the insert of a foreign DNA). Notice that in the construction of recombinant DNA, ligation between the vector DNA and the foreign DNA is always less than perfect. In the transformation step that follows, both the pUC vector and the recombinant vector DNA in the ligation reaction mixture will be picked up, and the host cells containing either type will be ampicillin resistant. These two types of transformants are distinguished by a selection method involving α-complementation.

In practice, the *E. coli* host of pUC is a mutant strain in which the α-peptide sequence region of the *lacZ* gene is missing. Hence, when the mutant *E. coli* harboring a pUC vector is plated on a growth medium containing IPTG, the α-peptide produced by the vector *lac Z'* gene will associate with the truncated β-galactosidase produced by the host to form a functional enzyme. This association is referred to as α-complementation. Functional β-galactosidase converts X-gal (5-bromo-4-chloro-3-indolyl-β-D-galactoside) to a blue product. Therefore, mutant *E. coli* harboring a pUC vector will grow on a medium containing IPTG and X-gal to give blue colonies. In the case that the host contains a recombinant vector, the enzyme will be non-functional, because the *lacZ'* region in the pUC vector is interrupted by the insertion of a foreign DNA. Hence, the colonies are white (Fig. 9.3).

Promoters and RNA Polymerases. The *lac* promoter described in the pUC system is commonly used for vector construction. The expression is inducible by IPTG, and transformants can be screened by the color of the colonies. In recent years, the bacteriophage T3, T7, and SP6 promoters are also used in the construction of bacterial expression vectors. These promoters are only recognized specifically by their respective RNA polymerases, and not by the *E. coli* RNA polymerases.

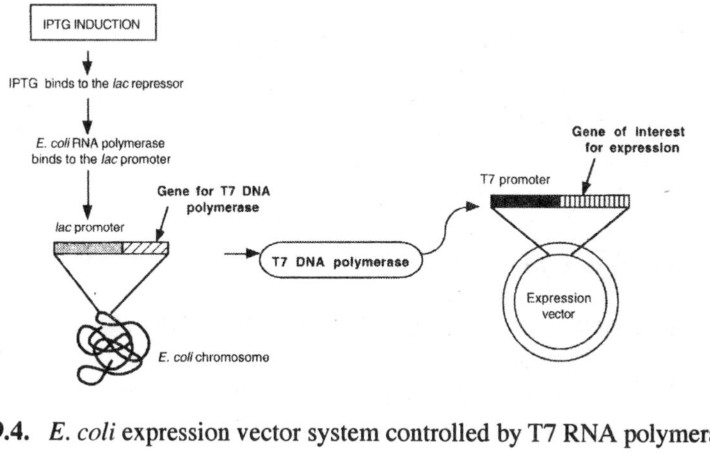

Fig. 9.4. *E. coli* expression vector system controlled by T7 RNA polymerase.

For example, in an expression system using the T7 promoter, the gene of interest inserted downstream under the control of this promoter will not be expressed until a source of T7 DNA polymerase is provided. The *E. coli* host used in this case contains a chromosomal copy of the T7 RNA polymerase gene under the control of a *lac* promoter inducible by IPTG. Therefore, the transformant will have the gene expressed, if IPTG is added to induce the production of T7 RNA polymerase, which in turn recognizes the T7 promoter in the vector to start transcription (Fig. 9.4).

Topoisomerase-based Cloning. The recent development of of topoisomerase-catalyzed systems provides an alternative to the use of ligase in joining DNA fragments. The biological function of topoisomerase is to cleave and rejoin DNA during replication. It has been found that recombination mediated by the vaccinia virus enzyme in *E. coli* is sequence specific. Binding and cleavage occur at a pentameric motif 5'-(C or T)CCTT in duplex DNA. The enzyme forms a complex between a tyrosine residue and the 3' phosphate of the cleaved DNA strand. The phospho-Tyr bond is then attacked by the 5'-OH of the original cleaved strand or of another donor DNA, resulting in religation and releasing the enzyme from the complex. This unique property of the enzyme has been harnessed for deriving a one-step strategy. The vector is constructed to contain CCCTT recognitions sites at both ends of the DNA when linearized. This enables the vector to ligate DNA sequences with compatible ends (Fig. 9.5). Both sticky end and blunt end ligations can be achieved. One can also place the cleavage site sequence on the insert and clone the DNA into a vector.

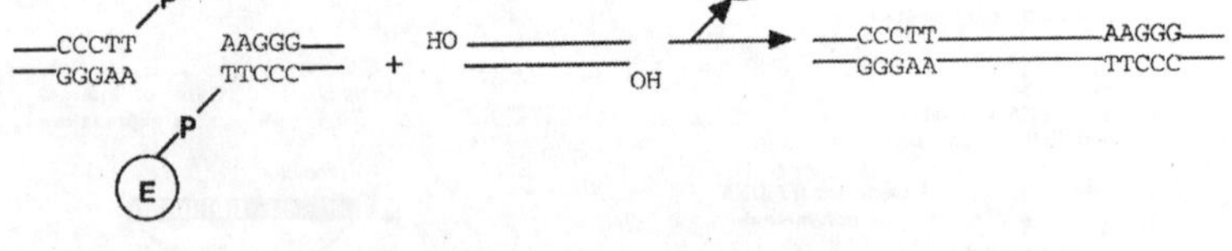

Fig. 9.5. Ligation by the use of topisomerase-catalyzed reaction.

In vitro transcription and translation. *In vitro* transcription/translation systems is a minimalist approach throwing out the cellular components not needed, and keeping those that are to be utilized for transcription and translation in a test tube. It can potentially eliminate lengthy cloning steps, and yields ng to ug of translated proteins (up to 500 μg in preparative scale systems) suitable for biochemical analysis. This procedure is often used as a quick way to screen proteins from a vast number of genes. Specialty vectors designed for the expression of cloned genes using *in vitro* systems are available commercially.

In the transcription step, the highly efficient bacteriophage RNA polymerase is used, with the RNA template containing a phage promoter, such T3, T7, or SP6 promotor. For bacterial genes, the mRNA used for the reactions needs a Shine-Dalgano sequence, and eukaryotic mRNA requires a 7-methylguanylate cap and a Kozak sequence for *in vitro* transcription (see Section 5.5.4). The *in vitro* translation system contains cell-free extracts from *E. coli* , rabbit reticulocytes, or wheat germ. The extracts contain ribosomal units, t-RNAs, aminoacyl-t-RNA synthetases, other requisite translation components, energy sources (ATP or GTP), amino acids, and cofactors. The choice of cell extracts depends on the gene source; *E. coli* extract is the frequent choice if prokaryotic genes are used. Transcription and translation can be performed as separate reactions, or simultaneously in one tube.

The strategy can be refined by additional modifications. (1) The gene product can be labeled by incorporating biotin-lysine-tRNA into the reaction. Labeling the translated enzyme would eliminate false positives in the following analysis and provide better interpretation of the enzyme activity. (2) The translated protein can be tagged with polyhistidine to facilitate purification and detection if desired for downstream analysis (see Section 9.1.1).

9.1.2 Bacteriophage Vectors

Another type of vector is derived from bacteriophage that includes bacteriophage λ and M13. The λ vectors and its derivatives are used mostly for the construction of cDNA or genomic libraries (see Section 11.1, 11.2). M13 vectors were used primarily to obtain ssDNA for DNA sequencing before the development of more recent sequencing methods (see Section 8.8).

Bacteriophage λ Life Cycle. Bacteriophages (abbreviated as phages) are viruses that infect bacterial cells. A virus exists as an infectious particle called virion in its extracellular phase. A phage λ particle has a head-and-tail structure, consisting of a core of DNA within a protein coat (capsid) that is joined to a helical protein structure (tail).

Phage λ is a temperate virus, in that its life cycle consists of two pathways - lytic and lysogenic. (Some phages show only lytic cycle, and these are called virulent phages.) In the infection of a bacterial cell, the phage λ particle is adsorbed on the cell membrane, followed by the injection of the λ DNA into the host cell. In the lysogenic pathway, the λ DNA is integrated into the bacterial genome. The integrated form of λDNA is called a prophage, and the host cell is now a lysogen. In the lytic mode, the λDNA integrates into the biosynthetic function of the host cell to produce more λDNA and proteins, which will be packaged into phage particles. To complete the cycle, the host cells are ruptured and the phages are released (Fig. 9.6).

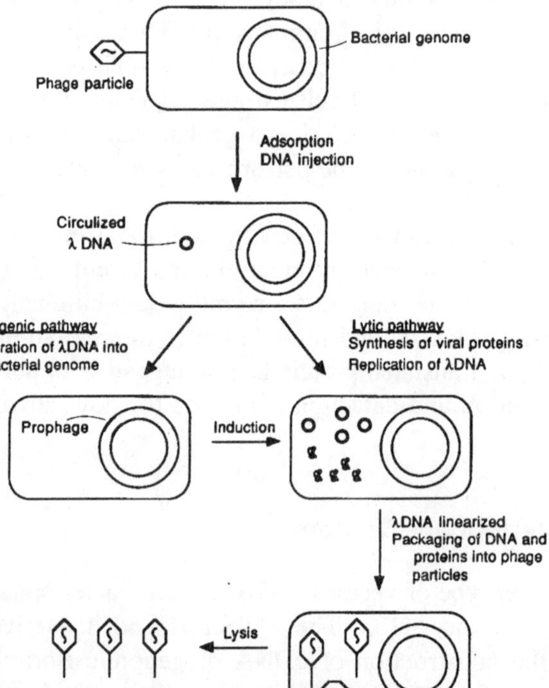

Fig. 9.6. Life cycle of bacteriophage λ.

Phage λ Vectors. Phage vectors used for recombinant work are designed to facilitate DNA insertion, screening for recombinants, and gene expression. Fig. 9.7A shows the physical map of λDNA in the intracellular circular form and in the linear form. Also presented is λgt11, an example of a phage vector (Fig. 9.7B).

(1) The λ vector contains a *lacZ* gene and a unique *Eco*RI restriction site at the 5' end of the gene. Non-recombinant phage grown on a bacteria lawn supplied with X-gal forms a blue plaque due to the hydrolysis of X-gal by β-galactosidase to a blue indolyl derivative (see Section

9.1.1). Insertion of a DNA segment or a gene at the unique restriction site interrupts the *lacZ* gene sequence. The β-galactosidase produced is inactive. Recombinant phages are recognized by the formation of clear plaques, which are distinguishable from the blue plaques for non-recombinants. The cloned DNA or gene sequence is expressed as a fusion protein with β-galactosidase. This implies that it can also be screened by immunodetection methods.

(2) In the λ vector the genes related to integration are deleted, and thus no induction is required to switch from lysogenic to the lytic mode. A region containing the terminator for RNA synthesis is deleted to ensure the expression of the S and R gene involved in cell lysis.

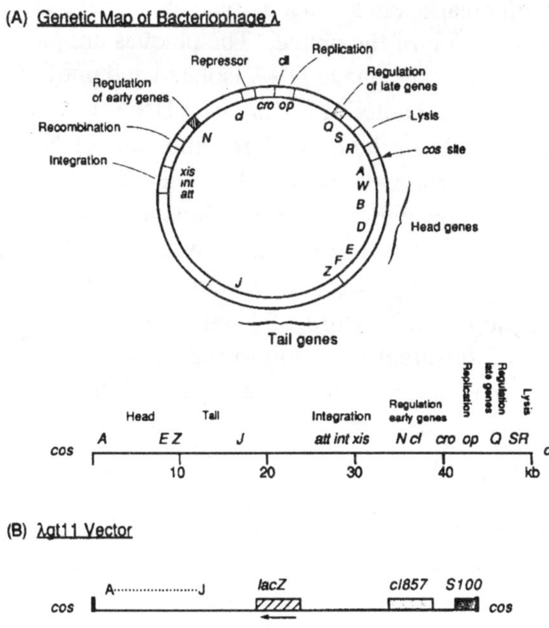

Fig. 9.7. Genetic map of bacteriophage λ and λgt11 vector.

(3) Amber (nonsense) mutations are introduced in the genes required for lytic growth. The mutations suppress the phage lytic function provided that a specific strain of *E. coli* capable of reversing the amber mutation is used as the host. This modification provides a safeguard against biological contamination of the environment. (Amber mutation is a point mutation that changes a codon into a stop codon. Consequently, the gene is expressed as an inactive protein with its carboxyl terminal seg-

ment missing. A reversion of this effect of mutation can be achieved by suppression in the anticodon of the tRNA carried out by the host strain.)

The total length of the recombinant DNA must be within the range of 75-105% of the normal λDNA genome (48.5 kb) to be efficiently assembled in the capsid during packaging. The size of λgt11 is 43.7 kb, and the vector can therefore accommodate up to 7.2 kb of insert DNA.

Transfection and In Vitro Packaging. Phage λDNA and recombinant λDNA can be introduced into host cells by mixing with a dense culture of competent ($CaCl_2$-treated) cells. (The introduction of viral DNA into a host cell is known as transfection.) The mixture is poured onto petri dish with appropriate growth medium. Incubation will result in a bacteria "lawn" dotted with clear spots known as plaques. These clear plaques are formed by the lytic action of the phage. The plaques are picked and grown in a suitable medium, and the phage DNA isolated and purified.

Transfection of bacterial cells by λDNA usually yields ~10^5 plaques per mg of DNA. In the case of recombinant λDNA, the yield decreases by one to two orders of magnitude. The efficiency will be greatly enhanced (10^7 - 10^8), however, if the recombinant λDNA is packaged into phage particles *in vitro*, allowing the recombinant λDNA introduced into the host cell by the natural process of infection. In practice, the recombinant λDNA is added to a mixture of lysates from two lysogen strains. Each strain carries a different mutation in the capsid (phage protein coat). Individually, the lysogens are unable to package λDNA and viral proteins into phage particles. A mixture of the two lysates, however, will render complementation of all the components necessary for packaging.

M13 Bacteriophage Life Cycle. Developed in parallel with the pUC series, the M13 vector system was widely employed for generating ssDNA for dideoxy DNA sequencing, before the development of newer sequencing methods. M13 is a filamentus bacteriophage of male *E. coli*.

M13 phage particle contains single-stranded circular DNA ((+) strand). Following infection of the *E. coli* host cell, the (+) strand serves as template for the synthesis of the complementary (-) strand. The double-stranded viral DNA in the *E. coli* host cell, referred to as the replicative form (RF), is replicated and amplified to 100-200 copies per cell. In a later stage, the (+) strand continues to be synthesized, and the (-) strand is prevented from replication. The accumulated (+) strands are packaged with the viral proteins to generate phage particles (Fig. 9.8). On a bacterial lawn, the growth of M13 does not give clear plagues. Instead, the plagues appear turbid, because M13 is non-lytic (no dissolution of the bacterial cell wall).

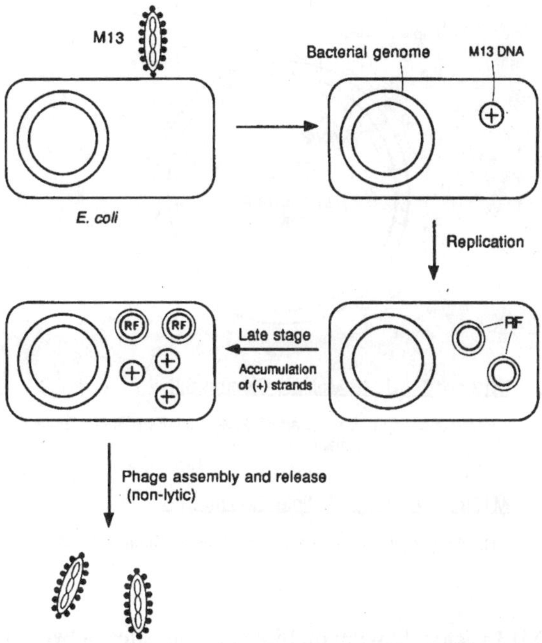

Fig. 9.8. Life cycle of M13 bacteriophage.

M13 Vectors. M13 vectors consist of modifications of the M13 wild type DNA: (1) A *lacI'OPZ'* operon (see Section 9.1.1) and (2) A multiple cloning site constructed in the *lacZ'* region (Fig. 9.9). It is essential for M13 infection that the *E. coli* host strain contains the F' episome (specialized plasmids containing an F factor that encodes sex pili in the male *E. coli* cells). The genes for proline biosynthesis are removed from the *E. coli* genome and inserted into the F' episome, so that the F' episome is retained when the strain is grown on minimal media (deficient in proline). The F' episome is also inserted with a mutated *lacZ* gene (lacking the α-peptide sequence), referred to as *lacZ*ΔM15, which is used for α-complementation, and a mutated *lacI* gene, referred to as *lacI^q*, which results in the overproduction of the *lac* repressor. Therefore, the genotype of the M13 host strain is *lacI^q*ZΔM15.

The M13 phage DNA is not infectious, but bacterial cells can pick up both ss and RF forms with $CaCl_2$ treatment in the same way as plasmids. The M13mp vectors are constructed in pairs (e.g. M13mp18 and M13mp19) with opposite orientation of the recognition sites in the multiple cloning site regions (Fig. 9.9). This enables cloning of a DNA fragment, in both orientation, and to be sequenced in both directions.

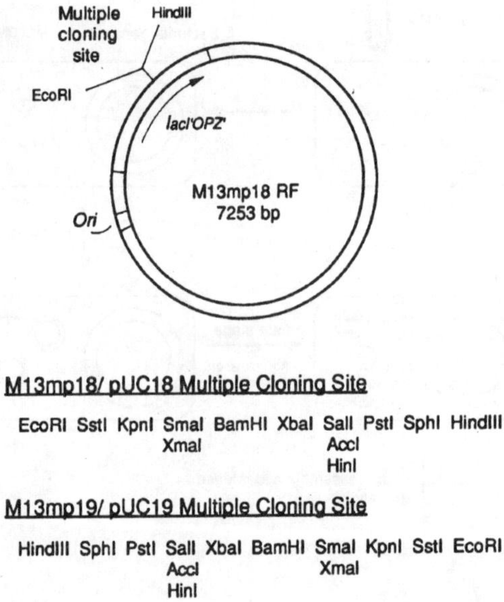

Fig. 9.9. M13 vector showing multiple cloning sites in two orientations.

9.1.3 Cosmids

Cosmids are plasmids containing a bacteriophage λ *cos* site. The hybrid structure enables insertion of large DNA fragments. The λ *cos* site is required for recognition in packaging (Fig. 9.10).

In the normal life cycle the λDNA molecules produced in replication are joined by the cohesive ends (*cos* site) to form a concatamer (long chains of DNA molecules). In the steps that follow, the concatamer is enzymatically cleaved, and each individual λDNA molecule is packaged into phage particles. Therefore, by incorporating a *cos* site into a plasmid, the resulting hybrid vector can be used for *in vitro* packaging. The size of a foreign DNA that can be inserted into the cosmid, and packaged into phage particles, is restricted to a range of 35-45 kb, assuming the cosmid size is ~5 kb. As a general rule, the total length of cosmid plus the DNA insert should be ~75-105% of the size of phage λDNA for efficient packaging. The phage particles are then used to infect *E. coli*. The recombinant cosmid DNA circularizes once inside the *E. coli* cell, and replicates like a plasmid.

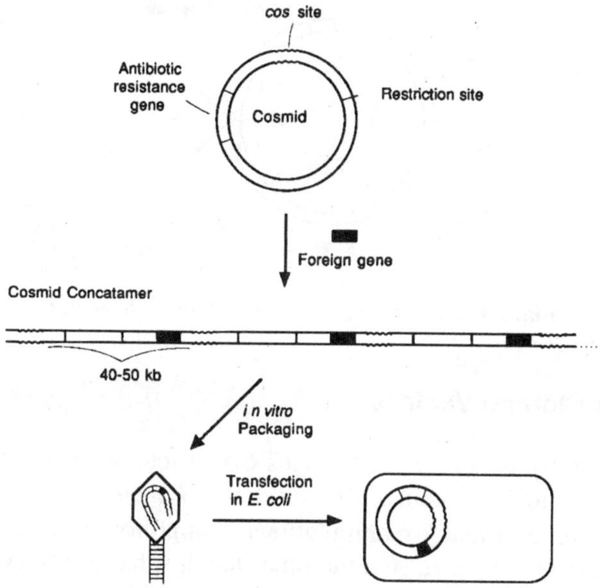

Fig. 9.10. Cosmid replicates like a plasmid and is packaged like phage λDNA.

9.1.4 Phagemids

Vectors have been designed to combine features from filamentous phage and plasmid. These "phagemids" allow the propagation of cloned DNA as conventional plasmids. When the vector-containing cells are infected with a helper phage, the mode of replication is changed to that of a phage in that copies of ssDNA are produced.

(1) A phagemid contains a bacterial plasmid origin of replication (e.g. ColE1 *ori*) and a selectable marker (e.g. ampicillin resistance gene) which enable propagation and selection in the plasmid form (Fig. 9.11).

(2) A filamentous phage origin of replication enables the production of ssDNA under the infection with a helper (filamentous) phage. The gene II protein expressed by the helper phage promotes single-stranded replication of the clone at the origin. The ssDNA is circularized, packaged, and released.

(3) A multiple cloning site inserted into the *lacZ* α peptide sequence, so that blue/white color selection can be used in screening for insert-containing clones. This construction also results in the expression of the DNA insert as a β-galactosidase fusion protein.

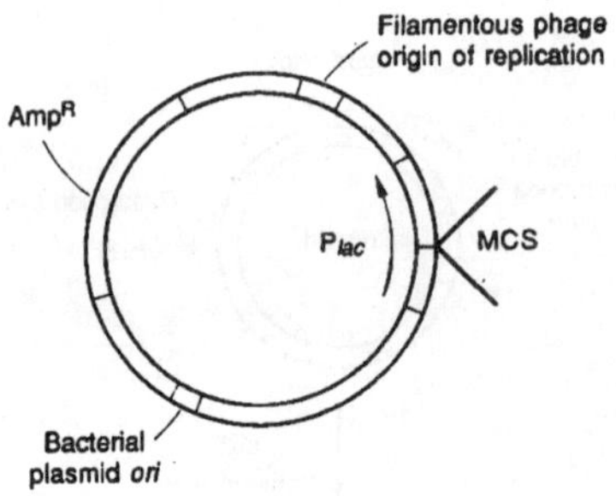

Fig. 9.11. Structural organization of a phagemid.

9.2 Yeast Cloning Vectors

Yeast offers several advantages over bacterial systems for the expression of complex proteins. The yeast *Saccharomyces* is a unicellular microorganism, and many manipulations commonly used in bacteria can be readily applied to yeast. On the other hand, it has a eukaryotic cellular organization, like those of plants and animals, making it a frequent choice as an appropriate host system for the production of proteins that may require posttranslational modification for full biological activity. Traditionally, the brewer's or baker's yeast, *Saccharomyces cerevisiae*, has been the biotechnologists' choice. A growing number of expression systems using other yeasts are becoming available, e.g. *Hansenula polymorpha*, *Pichia pastoris*, *Kluyveromyces lactis*.

9.2.1 The 2μ Circle

Yeast cloning vectors have been developed based on a plasmid, called 2μ circle, found in yeast. The 2μ circle is 6318 bp in size, and present in the nucleus of most *Saccharomyces* strains at ~60-100 copies.

(1) Yeast vectors contain the origin of replication from the 2μ circle (Fig. 9.12). Alternatively, the autonomously replicating sequence (ARS) from the yeast chromosomal DNA can be used. Vectors containing either the 2μ circle replication origin or the ARS are able to replicate in the yeast following transformation. Vectors that do not contain the 2μ circle replication origin or the ARS are called "integrative" vectors, because the vector DNA integrates into the yeast chromosome.

(2) A selectable marker for the screening of transformation in yeast is usually incorporated. Examples of frequently used markers are

LEU2 and *URA3*. The *LEU2* gene codes for β-isopropylmalate dehydrogenase, an enzyme involved in the synthesis of leucine. In a host system in which mutant yeast (that lacks the *LEU2* gene) is used, yeast transformants harboring the vector will grow on a medium that is deficient in leucine, whereas non-transformants will not survive. The selection is based on the complementation of the deficiency in the yeast host by the vector *LEU2* marker gene. Another approach is to use an essential gene such as *URA3* as a selectable marker. Mutant yeast strains (used as host) lacks the gene cannot synthesize uridine monophosphate, and will not survive even in a rich medium. Only transformants harboring the vector (with *URA3*) can grow.

A dominant marker, such as *CUP1* that confers resistance to copper, can be used. The *CUP1* marker bears positive selectable characteristics, and hence does not need for mutant yeast strain as host. This type of markers is practically useful when working with industrially important yeast strains (for example, strains used for brewing) that cannot be suitably mutated.

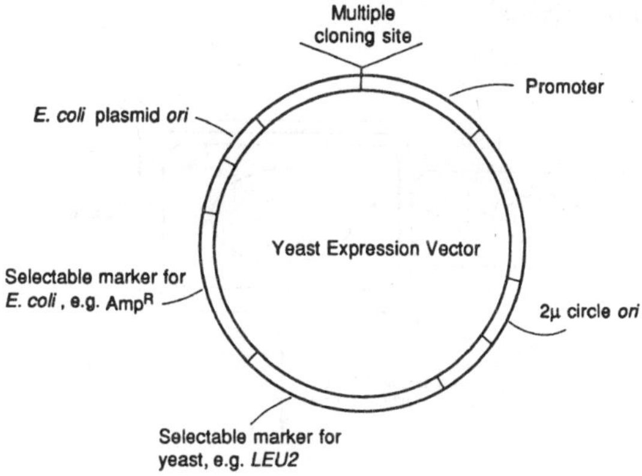

Fig. 9.12. Structural organization of a yeast expression vector.

(3) A suitable promoter is needed for gene expression. Two types of promoters are used: (A) for constitutive expression (i.e. The gene is expressed continuously during the culture of the yeast cells), and (B) for regulated expression (i.e. The gene is expressed in response to an external signal.) Constitutive expression becomes a problem working with gene products that are toxic to yeast for a number of reasons. (A) The growth rate of the yeast culture is low. (B) There is an unfavorable selection against cells expressing the gene product. (C) Consequently, the yield of the gene product is low. Using regulated expression vectors, expression can be switched after the yeast culture reaches a high cell density. An ex-

ample of highly regulated promoters is the GAL1 promoter, with expression induced 2000-fold by the addition of galactose.

Some yeast vectors also contain a replication origin from an *E. coli* plasmid (e.g. pBR322 *Ori*, ColE1 *Ori*), and a selectable marker that enables the vector to work in a bacterial host. This type of vectors that can operate in both yeast and *E. coli*, is called a "shuttle vector". Using a shuttle vector allows DNA manipulation to be conducted by conventional procedures in bacterial system, and the final gene construct can be placed in yeast for expression (see Section 17.2.3 for description for another type of yeast vector, yeast artificial chromosomes).

9.3 Vectors for Plant Cells

The Ti (tumor-inducing) plasmid is widely utilized to introduce DNA into plant cells. The Ti plasmid is isolated from *Agrobacterium tumefaciens*, a soil bacterium that infects plants, causing the formation of crown gall (tumor tissue).

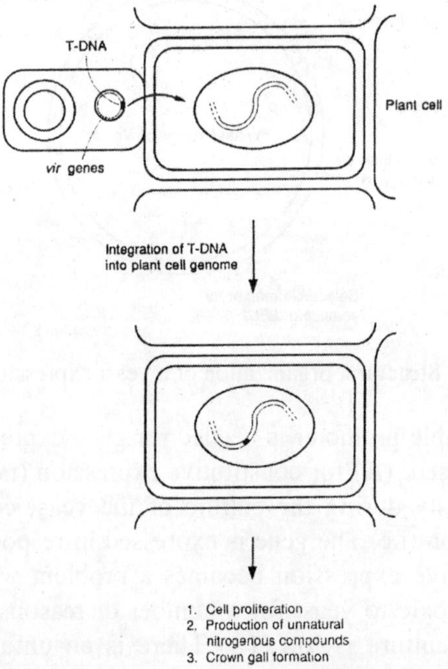

Fig. 9.13. Infection of plant cells by *Agrobacterium*.

In the infection process, a small (~20 kb) segment called T-DNA in the Ti plasmid is transferred and integrated into the plant chromosome. The transfer is controlled by the *vir* (virulence) gene located in the Ti plasmid (Fig. 9.13).

9.3.1 Binary Vector System

Ti plasmid in its natural form is not suitable as cloning vector for at least two reasons. (1) Plant cells infected with Ti plamid turn into tumor cells that cannot be regenerated into plants. (2) The size of Ti plasmid is 150-200 kb, making it extremely difficult to manipulate.

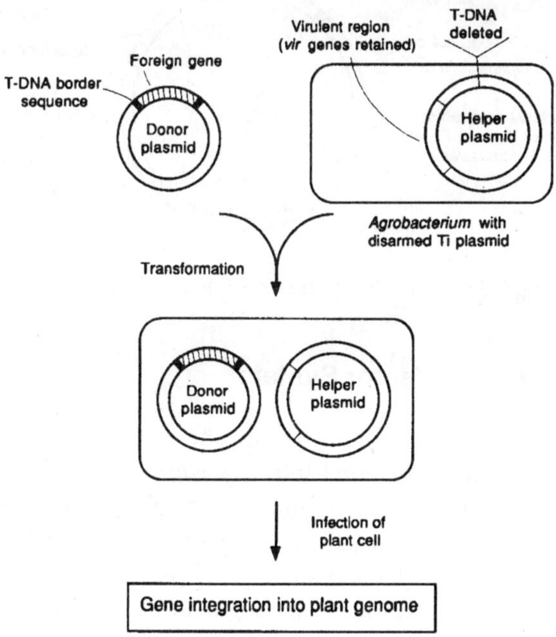

Fig. 9.14. Cloning strategy using a binary vector system.

For cloning use, a binary vector system is used, consisting of a helper plasmid and a donor plasmid. The helper plasmid is a "disarmed" Ti plasmid with the entire T-DNA (which carries the tumor-causing genes) deleted. A donor plasmid is a small *E. coli* plasmid, carrying a truncated T-DNA (25 bp border sequences of the intact T-DNA region) that contains the sites for excision. Insertion of a gene is made at the truncated T-DNA region. The two plasmids function in a complementary manner. The donor plasmid carries the inserted gene flanked by the border sequences of T-

DNA for excision. The helper plasmid provides the enzymes (coded by the *vir* genes and others) to direct transfer of the recombinant T-DNA (Fig. 9.14).

In practice, the *Agrobacterium* strain carries the helper plasmid (Ti plasmid disarmed). The donor plasmid, called the binary cloning vector, is a bacterial vector, consisting of (1) replication origins for both *E. coli* and *Agrobacterium*, (2) selectable markers for bacteria and for plants, (3) T-DNA border sequence of Ti plasmid, and (4) a cloning site for insertion of the gene of interest (Fig. 9.15).

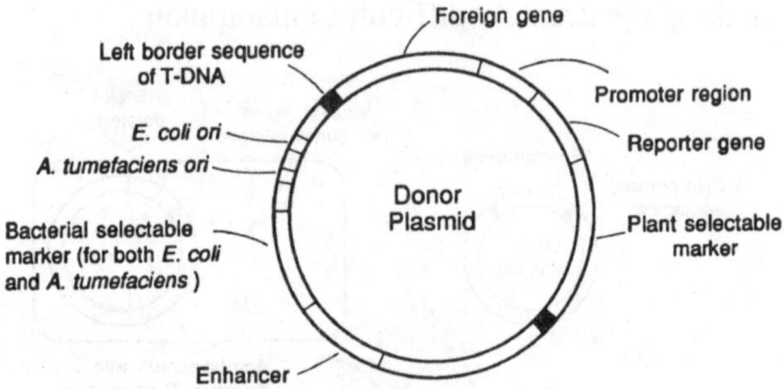

Fig. 9.15. Structural organization of a donor plasmid.

9.3.2 Cointegrative Vector System

In a cointegrative vector system, the gene that is to be introduced into the plant genome is inserted into a plasmid vector. The intermediate cloning vector contains: (1) a replication origin for *E. coli* (but not for *Agrobacterium*), (2) a plant selectable marker, (3) a bacterial selectable marker, (4) a T-DNA border sequence of Ti plasmid, (5) a sequence of Ti plasmid DNA homologous to a DNA segment in the disarmed Ti plasmid, and (6) a cloning site for insertion of the gene sequence.

Following cloning, the *E. coli* transformants are selected based on antibiotic resistance. The intermediate cloning vector carrying the gene is transferred from *the E. coli* to *Agrobacterium* containing a disarmed Ti plasmid by mating. Once in the *Agrobacterium*, the gene sequence is integrated into the disarmed Ti plasmid by recombination, because both the vector and the Ti plasmid contain a homologous short sequence (Fig. 9.16). As the intermediate vector lacks the replication origin for *Agrobacterium*, those that do not integrate will not accumulate. The recombinant *Agrobacterium* is identified by the bacterial selectable marker, and used to infect plant cells. This is done by inoculating cotyledon explants with the recombinant *Agrobacterium*. The transformed explant carries the intermediate vector, and therefore the plant selective marker for screening of the plant cell transformants.

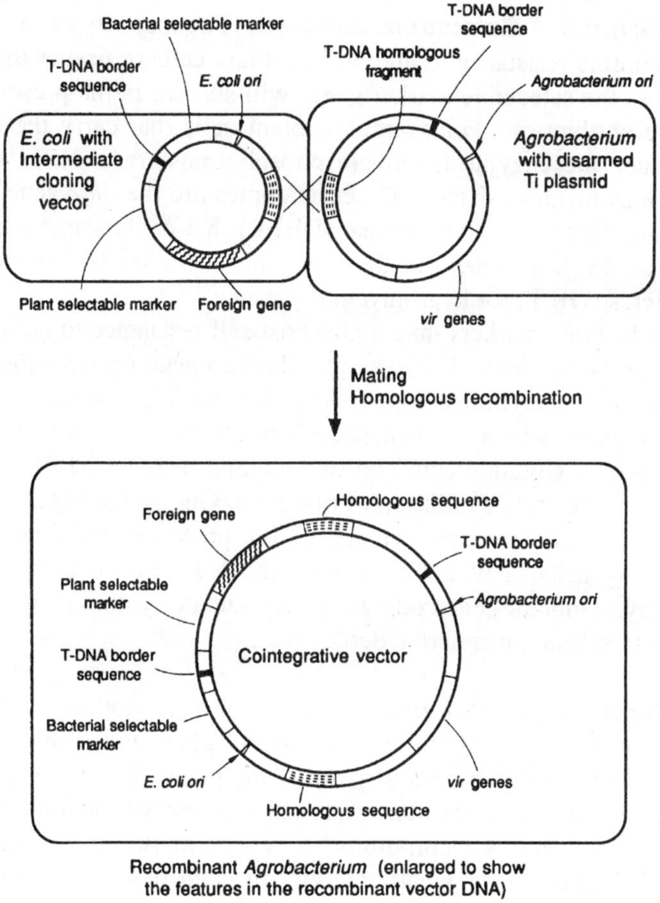

Fig. 9.16. Cloning strategy using a cointegrative vector.

9.3.3 Genetic Markers

The use of binary vector system or integrative vector system needs two types of genetic markers. A bacterial selectable marker is needed for selecting *E. coli* transformants in the manipulation of gene constructs. A second marker for transformed plant cells is also needed. Genetic markers can be grouped into two categories. (1) Dominant selectable markers are genes encoding a product that allows the cells carrying the gene to grow under specific conditions so that the transformants can be selected. (2) Screenable markers are genes that encode a product that can be rapidly detected.

Dominant Selectable Markers. The majority of this group of markers confers resistance to antibiotics. Plant cells or tissues that contain and express the selectable marker genes will survive in the presence of the respective antibiotics. For examples, plant cells that carry the *neo* gene will produce the enzyme neomycin phosphotransferase (NPTII), making the cells kanamycin resistant. Other examples are the *dhfr* gene encoding the enzyme dihydrofolate reductase (DHFR) that confers methotrexate resistance to the host cell, and the *hpt* gene encoding hygromycin phosphotransferase (HPT) for hygromycin resistance.

Selectable markers that confer host cell resistance to herhicides are increasingly being used. For example, the *bar* gene encodes the enzyme, phophinothricin acetyltransferase (PAT), that confers resistance to the herbicide, bialaphos, which is a tripeptide consisting of phosphinotricin (PPT) (an analog of a glutamic acid and two alanine residues). PPT is a potent inhibitor of glutamate synthase, a critical enzyme in the regulation of nitrogen metabolism in plants. Hence, in the presence of the herbicide bialaphos, only cells or plants containing the *bar* gene will survive. In a similar way, a mutant gene coding for acetolactate synthase (ALS) confers resistance to sulfonylurea herbicides.

Screenable Markers. In plant cells, transcriptional activity may vary and interact with subtle environmental changes. Some promoters have localized activity in various parts of the plant. It is sometimes desirable to run rapid testing for transcriptional regulatory functions of promoters and/or enhancers by incorporating a genetic marker that enables histochemical detection of enzymatic activity in plant tissues. These markers are sometimes called reporter genes, because they report the biochemical activity of certain targeted genetic elements in the plant cells or tissues or whole plants. In contrast to dominant selectable markers, these markers do not facilitate screening of transformed cells for survival under specific conditions. Rather, reporter genes service to tag transformed cells for the purpose of investigating transient gene expression or to establish transformation and transgenic plants.

Examples include the *gus*, the *luc*, and the *cat* genes. The *gus* gene from *E. coli* encodes the enzyme β-glucuronidase, which breaks down histochemical substrates, such as 5-bromo-4-chloro-3-indolyl β-D-glucuronide into a blue color compound. A fusion of the *gus* gene with the promoter allows spatial visualization of gene expression, and hence a detail analysis of the cell-specific expression directed by transcriptional activities of individual promoters.

The luciferase (*luc*) gene from firefly encodes an enzyme that catalyzes a light-producing reaction in the presence of adenosine triphosphate (ATP), oxygen, and luciferin (a substrate). Transgenic plants or transformed plant cells carrying the *luc* gene can be rapidly selected by simple detection of luminescence. It is widely used as a reporter gene of gene expression, genetic crosses, and cell functions.

The bacterial *cat* gene (coding for chloroamphenicol acetyltransferase (CAT) and the *lacZ* gene (coding for β-galactosidase) are common alternatives. The CAT protein catalyzes the acetylation of chloroamphenicol, and β-galactosidase cleaves the β-1,4-linkage in a glucan substrate.

Selected Examples of Genetic Markers

Gene	Protein	
Dominant Selectable Markers		**Resistance**
neo	neomycin phosphotransferase II (NPTII)	kanamycin
dhfr	dihydrofolate reductase (DHFR)	methotrexate
hpt	hygromycin phosphotransferase (HPT)	hygromycin
bar	phosphinothricin acetyltransferase (PAT)	bialaphos (phosphinotricin)
als(mutated)	acetolactate synthase (ALS)	sulfonylurea
Screenable Markers		**Detection**
gus	β-glucuronidase (GUS)	colorimetric
luc	luciferase (LUC)	luminescent
cat	chloramphenicol acetyltransferase (CAT)	radioactivity
lacZ	β-galactosidase	colorimetric, flurometric, chemiluminescent

9.3.4 Plant Specific Promoters

Promoters used in plant cells are derived from either pathogens or plant gene promoters. An example of plant specific promoters obtained from pathogens is the cauliflower mosaic virus (CaMV) 35S promoter.

Transcription of genes controlled by the CaMV35S promoter is generally considered to be constitutive (i.e. all-time expression) in various tissues of transgenic plants of a wide variety of species. The 35S promoter carries a highly efficient enhancer (see Section 5.5.1). Promoters derived from plant genes are frequently tissue-specific, and regulated by environmental factors such as light and temperature. An example of this type of promoters is the *cab* promoter for the *cab* gene encoding the major chlorophyll a/b-binding protein. The *cab* promoter is light inducible.

9.4 Vectors for Mammalian Cells

Genetically engineered animal cell lines are useful for the production of human therapeutic proteins, and also provides a convenient system for studying gene regulation and control in eukaryotic cellular processes. There are, in general, two types of methods for transferring DNA into mammalian cells: (1) mediated by virus infection, or (2) transfection with mammalian expression vectors.

Viral-mediated transfer provides a convenient and efficient means of introducing eukaryotic genes into mammalian cells. This method involves the use of a number of viruses, such as simian virus 40 (SV40), bovine papilloma virus (BPV), Epstein-Barr virus (EBV), and retrovirus. Baculovirus is also included, although insect cells are used as the host in this system. It is not necessary, however, to use a viral vector to express foreign genes in animal cells, particularly if transient expression (several days to weeks) is desired. Mammalian expression vectors for this purpose are derived from plasmid DNA carrying regulatory sequence from viruses.

9.4.1. SV40 Viral Vectors

SV40 virus is one of the most studied Papovaviruses, with a genomic size of ~5 kb. It consists of 2 promoters that regulate early genes (encoding large T and small t antigens), and late genes (encoding viral capsid proteins VP1, VP2, and VP3). The SV40 virus also contains a replication origin that supports autonomous replication in the presence of the large T antigen.

Vectors are constructed by cloning the SV40 sequence containing the replication origin and the later promoter into a bacterial plasmid (e.g pBR322). The inserted foreign DNA replaces the viral late genes. The replacement of SV40 recombinant DNA as bacterial plasmid DNA provides an efficient means of DNA manipulation (Fig. 9.17).

After the proper construction of the SV40 recombinant DNA, the plasmid sequence is excised. The viral segment of the SV40 recombinant DNA is ligated and used for cotransfection of animal cells with a helper virus. A helper virus is a SV40 virus with defective early genes but has functional late genes to complement the viral recombinant DNA (in which the late genes are replaced by the foreign DNA). Host cells cotransfected by a recombinant DNA and a helper virus, therefore, are able to generate the viral DNA and all viral proteins necessary for packaging into infectious viral particles.

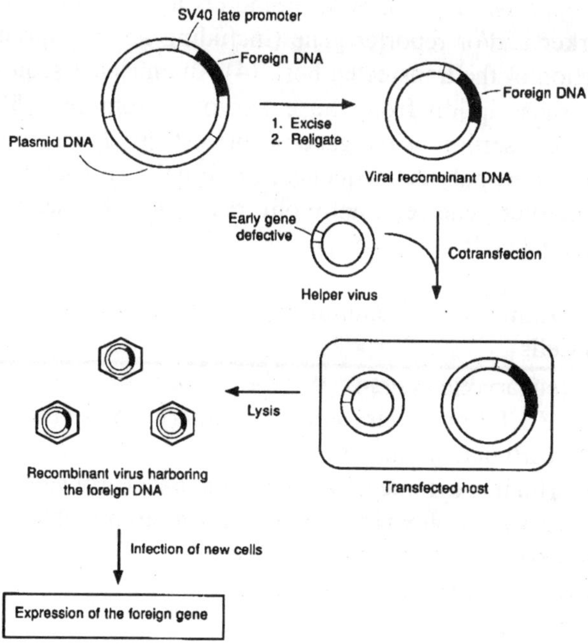

Fig. 9.17. Cloning strategy using a SV40 viral vector.

The use of a helper virus is obliterated if the host cell provides the viral functions instead. One such cell line, called COS, consists of African green monkey kidney CV-1 cells transfected with SV40 mutant virus with defective replication origin. Thus, the cell contains SV40 viral DNA integrated in its chromosomal DNA that can complement viral functions, but is incapable of replication. Several disadvantages limit the use of SV40 viral vectors: (1) The method is limited to applications using only monkey cells; (2) The expression is unstable due to cell lysis; (3) DNA rearrangement occurs during replication.

9.4.2 Direct DNA Transfer

For transient expression of transfected DNA in mammalian cells, alternatives to the use of viral infection are available. DNA can be introduced directly into mammalian cells (many cases, COS cells), by coprecipitation with calcium phosphate, electroporation, and other methods. A variety of transient expression vectors are commercially available for this purpose. Mammalian expression vectors typically include several structural features: (1) A replication origin for efficient amplification in mammalian cells (e.g. SV40 ori for COS cells, see Section 9.4.1); (2) An eukaryotic promoter for transcriptional regulation of the foreign gene targeted for expression (Viral gene promoters are usually used.); (3) A selectable marker and/or reporter gene (including an appropriate promoter) for the selection of the transfected host; (4) An enhancer sequence that acts to increase transcription from the eukaryotic promoter; (5) A multiple cloning site for insertion of the gene of interest; (6) Transcription termination sequence, and poly(A) sequence; (7) Finally, a bacterial replication origin, and marker gene (e.g. antibiotic resistance) for selecting transformants in bacterial cells.

Selected Examples of Common Features in Expression Vectors for Mammalian Cells

Promoters (eukaryotic system)	
MMTV-LTR promoter (mouse mammary tumor virus)	
SV40 early/late promoter	
CMV (human cytomegalovirus) immediate early gene promoter	
KT (herpes simplex virus thymidine kinase) promoter	
RSV (Rous sarcoma virus) promoter	
adenovirus major later promoter	
Selectable markers	
neo	aminoglucoside phosphotransferase
pac	puromycin acetyltransferase
hyg	hygromycin phosphotransfers
Screenable Markers	
cat	chloramphenicol acetytransferase
luc	luciferase
lacZ	β-galactosidase

Many promoters used in transient expression vectors are viral promoters, and can be either constitutive or inducible. SV40, RSV, CMV are examples of constitutive promoters for high-level transcription. On the other hand, MMTV-LTR promoter is inducible by glucocorticoids, steroid hormones that bind to receptors in the cells. The resulting hormone-receptor complex binds at DNA specific sites, resulting in the activation of transcription. Inducible promoters are useful in cases where the protein expressed by the cloned gene is toxic to the host cell.

Transfection of DNA using mammalian expression vectors is primarily transient, but approximately one in 10^4 cells will contain the foreign DNA in a stable integrated form. The use of dominant selectable markers enables the screening for stable DNA transfection and the generation of stable cell lines. Selectable markers usually employ antibiotic resistance genes, such as the *neo* gene that confers resistance to neomycin in bacteria and to G418 in mammalian cells. The *hyg* gene encoding hygromycin phosphotransferase, confers resistance to hygromycin. The *pac* gene from *Streptomyces alboniger* encodes an enzyme, puromycin acetyltransferase (PAC) that catalyzes N-acetylation of puromycin, making the antibiotic unreactive in mimicking aminoacyl-tRNA.

In addition to the above dominant selectable markers, other markers have also been used in a limited extent. These include hypoxanthine-gluanine phosphoribosyltransferase (HPRT), thymidine kinase (TK), and dihydrofolate reductase (DHFR), all utilizing specific enzyme activity as a tag. The use of this type of markers requires cell lines that are deficient in the corresponding enzymes.

9.4.3 Insect Baculovirus

Baculovirus expression systems have found increasing applications for the production of eukaryotic biologically active proteins. The system is similar to mammalian cells, in that it exhibits posttranslational processing - folding, disulfide formation, glycosylation, phosphorylation, and signal peptide cleavage (see Section 3.4). The system utilizes the baculovirus, *Autographa californica* multiple nuclear polyhedrosis virus (AcMNPV), which infects many species of *Lepidoteran* insects. The insect cells used in most laboratory experiments are derived from cultured ovarian cells of *Spodoptera frugiperda*.

Life Cycle of AcMNPV. Two viral forms exist: (1) extracellular virus particles, (2) virus particles embedded in proteinaceous occlusion. Viral occlusion is called a polyhedron. Proteins that form the occlusion are produced by the virus particle in the infected insect cell, and are called polyhedrin proteins. The life cycle begins with ingestion of polyhedron-contaminated food by a susceptible insect. The polyhedron entering the gut of the insect dissociates, releasing the virus particle, which infects the cells in the gut. Once inside the nucleus of the host cell, the virus particle replicates, and synthesizes viral proteins using the biological system of the host cell. The viral DNA and proteins assemble into new viruses, which are released from the cell by budding and capable of infecting other cells. In the later stage of infection, the virus particles convert into occlusions. The cell accumulates increasing numbers of polyhedron, and eventually lyses, releasing large number of polyhedron to the immediate environment.

Extracellular virus particles are responsible for cell-to-cell infection, while polyhdron is responsible for horizontal transmission of virus among insects. In other words, the gene for the polyhydrin protein is not essential for the production of virus particles in the cell, but only functional in the later stage for the production of polyhedron.

Baculovirus Transfer Vector. In practice, AcMNPV genome is too large (128 kb) to work with. A baculovirus transfer vector has to be constructed for cloning use. Transfer vectors contain: (1) a ~7 kb fragment of AcMNPV carrying the polyphedrin gene, and (2) a multiple cloning site constructed downstream of the polyhedrin gene promoter (Fig. 9.18).

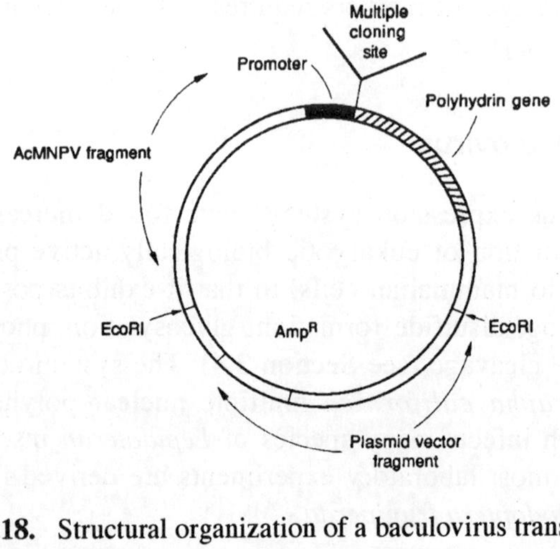

Fig. 9.18. Structural organization of a baculovirus transfer vector.

The gene of interest is to be inserted in the MCS. Both the recombinant transfer vector DNA and wild-type viral DNA are used to transfect insect cells. Within the cell, the inserted gene sequence is transferred to the AcMNPV viral DNA by homologous recombination forming the recombinant baculovirus DNA. Since insertion of a foreign gene at the MCS downstream of the polyhedrin gene promoter causes inactivation of the polyhedrin gene, cells carrying the recombinant baculovirus will be occlusion negative, visually distinguishable from cells containing occlusion positive wild-type virus (Fig. 9.19).

The frequency of recombination by this technique is less than 1%, and occlusion-negative plaques are frequently obscured among the high background of wild-type (occlusion-positive) plaques.

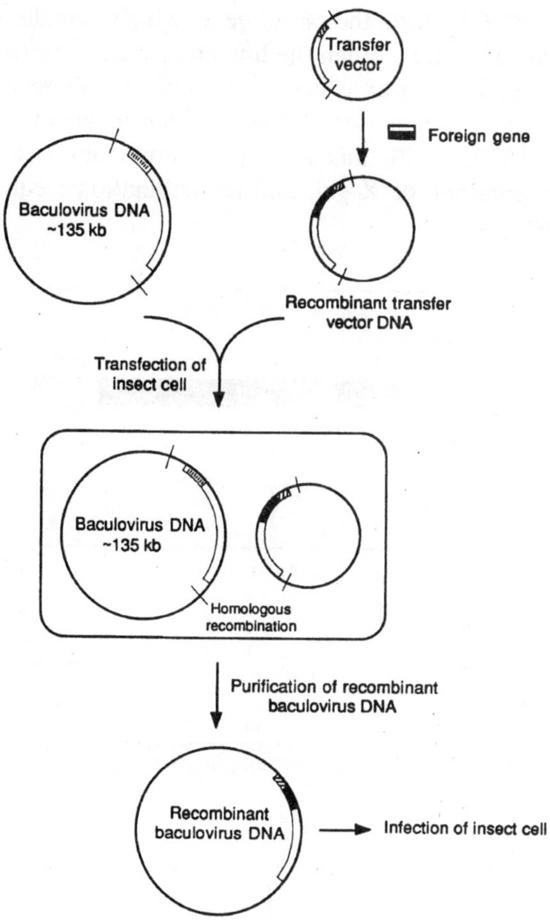

Fig. 9.19. Cloning strategy using a baculovirus transfer vector.

A more effective strategy has been developed based on the use of the parental viral DNA that incorporates a lethal deletion (Fig. 9.20).

(1) First, the AcNPV genome is modified to contain a truncated *lacZ* gene sequence upstream of the polyhedrin gene. It is further constructed with two *Bsu*36I restriction sites franking the polyhedrin locus –

one at the 5' end of the polyhedrin gene, and one within ORF1629 (coding for a capsid protein that is essential for viral viability). Digestion of this modified DNA with *Bsu*36I will linearize the DNA, with the release of the polyhedrin gene and part of the ORF1629 sequence. Since the ORF1629 gene is disrupted, insect cells infected with this linearized DNA cannot produce viable viruses.

(2) A transfer vector is constructed with sequences of intact *lacZ* gene and ORF1629 flanking the target gene, which is under control of the polyhedrin gene promoter. When the linearized bacculovirus DNA and the transfer vector are used to transfect insect cells, double recombination occurs, resulting in a circular viral DNA with the regeneration of the *lacZ* gene and ORF1629. With this system, recombinant viruses form blue plaques in the presence of X-gal, and recombination frequencies can be greater than 90%.

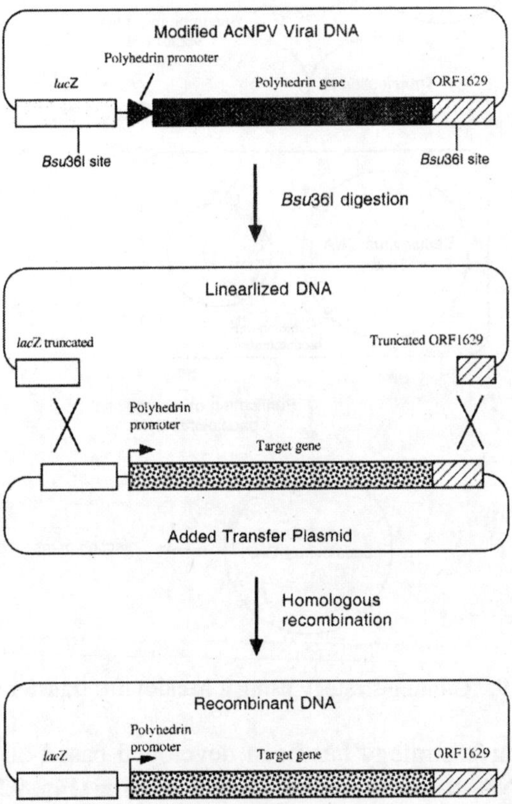

Fig. 9.20. Modified AcNPV vector.

9.4.4 Retrovirus

Retroviruses contain RNA as the genetic material in a protein core enclosed by an outer envelop. The viral RNA genome contains at the 5' and 3' ends, long terminal repeats (LTR) carrying the transcriptional initiation and termination, respectively. In between the 5' and 3' LTR regions, are three coding regions for viral proteins (*gag* for viral core proteins, *pol* for the enzyme reverse transcriptase, and *env* for the envelop), and a psi (ψ) region carrying signals for directing the assembly of RNA in forming virus particles (Fig. 9.21).

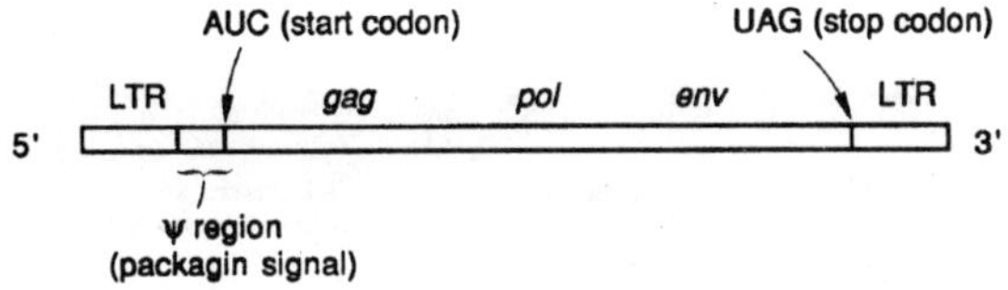

Fig. 9.21. Retroviral RNA genome.

In the infection process, the viral RNA released into the host cell is reverse transcribed into DNA strands that subsequently integrated into the host cell genome (Fig. 9.22)

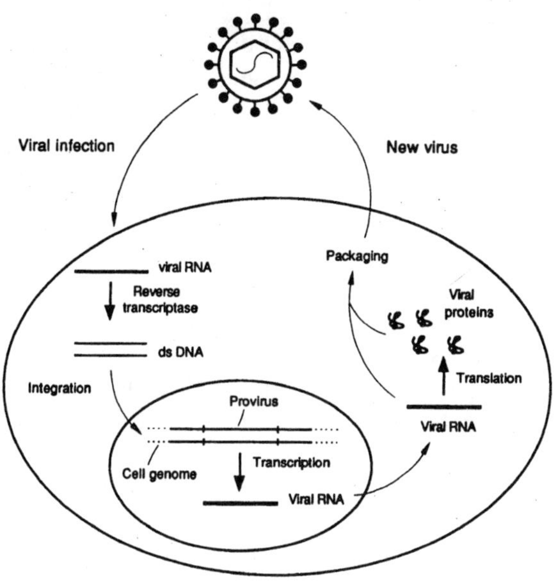

Fig. 9.22. Life cycle of retrovirus.

The integrated viral DNA, known as provirus, contains all the sequences for the synthesis of viral RNA and viral proteins. The integrated viral DNA is transcribed together with the cellular biosynthetic process. The transcribed viral RNA also serves as mRNA for the synthesis of viral proteins. The viral RNA and proteins are assembled in a process called packaging to generate new retroviruses.

Retrovirus Vector and Packaging Cell. Retroviruses cannot be used directly as vectors because they are infectious. Safe retrovirus vectors are constructed using a system consisting of two components: (1) retrovirus vector, and (2) packaging cell.

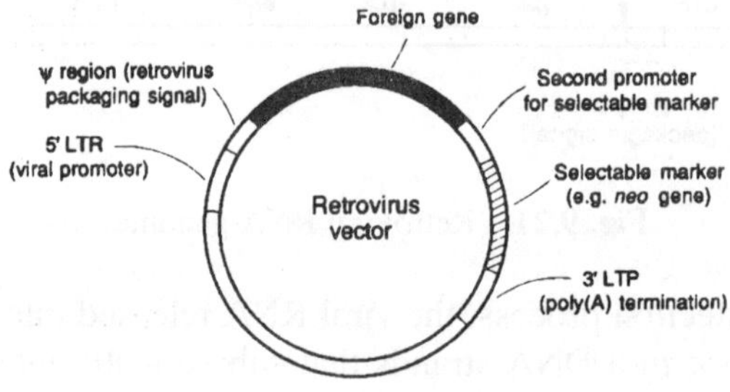

Fig. 9.23. Structural organization of retrovirus vector.

A retrovirus vector is a recombinant plasmid carrying a sequence of the viral genome (Fig. 9.23). In the construction of viral vector, most of the viral structural genes are deleted, but the LTR and psi (ψ) region are retained. The viral sequence is constructed with selective markers, such as the *neo* gene (karamycin resistance), or *hyg* gene (hygromycin resistance). The retrovirus LTR (long terminal repeats) strong promoter is usually used for the expression of inserted gene. Other promoters such as simian virus 40 (SV40) early promoter can also be used. A second promoter is used to control the expression of selective markers. A unique restriction site is constructed for insertion of foreign DNA

The vector containing the inserted gene is introduced into packaging cells by transfection. Packaging cells are derived from murine or avian fibroblast lines containing integrated provirus DNA with the ψ region deleted. During normal cell transcription and translation, the integrated provirus DNA provides the proteins (encoded by *gag*, *pol*, and *env*) required for assembly into viral particles for packaging, whereas the integrated recombinant vector DNA provides the RNA to be packaged (Fig. 9.24). The resulting virus particle is a safe vector that contains no viral proteins and cannot produce progeny. These safe vectors are used for cell infection.

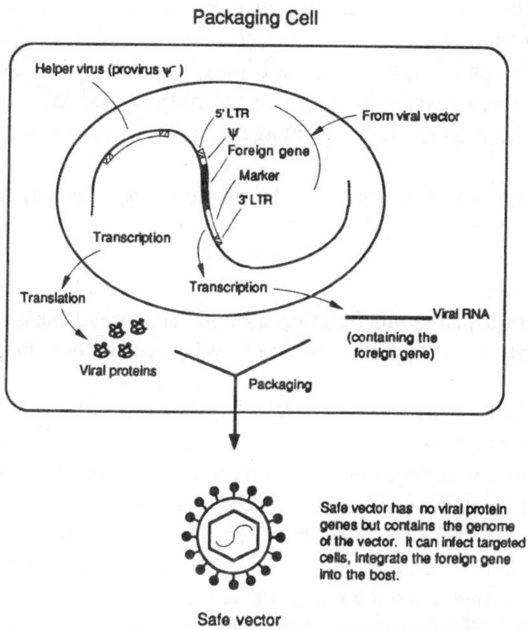

Fig. 9.24. Production of retrovirus safe vector.

Review

1. Describe the primary functions of: (A) replication origin, (B) antibiotic resistance gene, (C) multiple cloning site, in a plasmid vector.
2. What are selective markers? Give an example of a selective marker used in plasmid vectors and how it works.
3.

Function in *lac* operon	Function in pUC plasmid
lacI	*lacI'*
lacOP	*lacOP*
lacZ	*lacZ'*

4. What are the advantages of using bacteriophage T7 promoter in *E. coli* expression vectors?
5. How does topisomerase replace the use of ligase?
6. What are the features needed for *in vitro* transcription and translation? Describe their functions.
7. Describe the phage lytic cycle and the lysogenic cycle. List the similarities and differences between the life cycles of phage λ and M13.
8. What are the modifications of phage λ DNA in the construction of a phage vector.
9. What are the modifications of M13 phage in the construction of a M13 vector?
10. What are the unique features and properties of a cosmid that make it desirable as a cloning vector?
11. What makes a phagemid function as both phage and plasmid? What are the requirements for this dual function? Why is a helper phage needed in this system?
12. Describe the functions of (A) *ARS*, (B) *LEU2*, (C) *CUPI*, and (D) *URA3* in yeast cloning vectors.
13. List the structural features in a binary cloning vector, and a cointegrative vector. What are the similarities and differences?
14. What is the major distinction between dominant selective markers and screenable markers? Give examples of each marker type, listing the genes, the proteins (enzymes), and the nature of resistance or detection methods.
15. Do you find differences by comparing genetic markers used in vectors for plant and mammalian systems?
16. Why do baculovirus transfer vectors contain the polyhedrin gene sequence? Describe the advantages of using baculovirus expression systems in cloning.
17. Describe the major stages in the life cycle of a retrovirus.
18. Explain the purpose of the following steps taken in the construction of a retrovirus vector.
    (A) deletion of viral structural genes
    (B) retainment of LTR and Ψ region
    (C) insertion of a selectable marker
    (D) insertion of promoters
19. What is a retrovirus safe vector? What are the key steps in producing a safe vector?

CHAPTER 10

TRANSFORMATION

After insertion of a foreign DNA into a vector, the next step is to introduce the resulting recombinant DNA into a suitable host cell. The process of introducing DNA into living cells is called transformation. In a special case where the introduced DNA is a viral DNA, the process is called transfection. The choice of methods depends on the type of host systems in use, as well as the objective of cloning. Some of the procedures for transformation have been briefly mentioned in the discussion of vectors. A more detailed discussion on this important process is presented in this chapter.

10.1 Calcium Salt Treatment

A foreign DNA can be easily introduced into bacterial cells if the cells have been pretreated with $CaCl_2$ or a combination with other salts. The treated cells are called competent cells that are able to take up DNA readily. With other cell types, transformation generally requires additional treatments. Yeast, fungal, and plant cells contain cell walls in some cases need to be digested to produce protoplasts (cell minus cell wall) before DNA can be picked up.

A common method for introducing a foreign DNA into mammalian cells involves coprecipitation of the DNA with calcium phosphate (mixing the purified DNA with buffers containing calcium chloride and sodium phosphate), and the mixture is presented to the cells in the cultured medium. The individual DNA usually integrates as multiply copies in the cell genome (see Sections 18.1 and 19.1). A wide selection of competent cells featuring a full range of cloning efficiencies and genotypes is commercially available.

10.2 Electroporation

To increase the efficiency of DNA uptake, electroporation is frequently used. The procedure is employed in yeast, fungi, and plant cells, and less frequently in bacterial systems and animals. In this procedure, the cells are subjected to a brief electrical pulse, which causes a localized transient disorganization and breakdown of the cell membrane, making it permeable to the diffusion of DNA molecules. The vector DNA (carrying a foreign DNA or not) can then be picked up by the cells.

10.3 *Agrobacterium* Infection

The use of Ti plasmid vectors for plant cells has been discussed (see Section 9.3). In practice, transformation is achieved by providing the *Agrobacterium* (carrying either cointegrative or binary vector) with wound cells. (1) Explant inoculation involves the incubation of sectioned plant tissues (leaf, stem, tuber, etc.) with the bacterium, culturing on medium for the growth of callus. Shoots and roots are then induced to grow by subculturing the callus in appropriate medium. (2) Protoplast co-cultivation employs isolated protoplasts with partially regenerated cells incubated with the bacterium. This is followed by culturing and subculturing the cells into callus, shoots and roots. (3) Seedling inoculation involves inoculating imbibed seeds with the bacterium. Transformants can be selected at the initiation of callus growth, at later stages as well as in transgenic plants. Explant inoculation is the most widely used procedure. Protoplast co-cultivation and seedling inoculation are generally limited only to certain species.

10.4 The Biolistic Process

Direct transfer of DNA into plant cells (also used in other types of cells) can be achieved by using the biolistic process. This is a direct physical method to transform cells *in situ*. In this process, a thin coat of DNA is deposited onto the surface of 0.5-1.5 μm tungston or gold microbeads. The DNA-coated beads are then loaded and fired from a "gene gun" by explosive, electric, or pressure charge. The DNA-coated beads are bombarded onto plant tissues, enter the cells, and are integrated into the cell chromosomal DNA randomly (Fig. 10.1). In the case of plants, the cells are regenerated into plantlets by tissue culture technique and grown into full plants. Although originally developed for plant cells, this method can be applied to animal cells, tissues, and organelles, yeast, bacteria and other microbes.

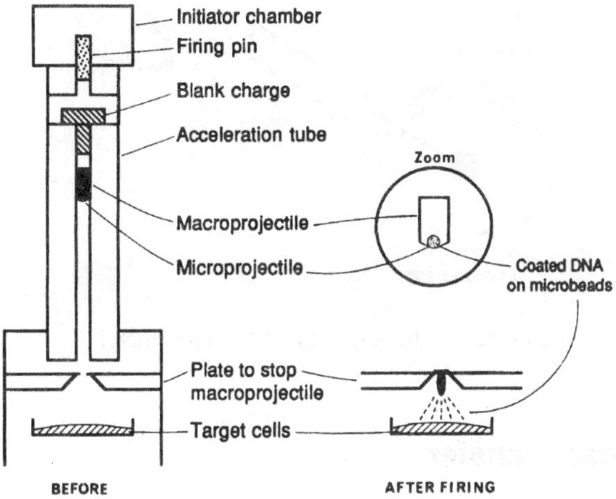

Fig. 10.1. Gene gun for firing coated DNA beads into plant cells.

10.5 Viral Transfection

This has been discussed in relation to the construction of retrovirus vectors for animal cells, and bacteriophage λ for bacterial cells (see Section 9.4).

10.6 Microinjection

Transformed plant cells can be regenerated into transgenic plants carrying the cloned DNA. However, animal cells cannot be regenerated into transgenic animals. For the production of transgenic animals, the DNA is injected into the pronuclei of the fertilized egg using a micropipet (Fig. 10.2). For expression purposes, the gene of interest must be properly constructed with a promoter region and other control elements to direct tissue-specific production of the protein. The transformed zygote is implanted into a surrogate mother to give birth to transgenic offspring (see Section 21.1)

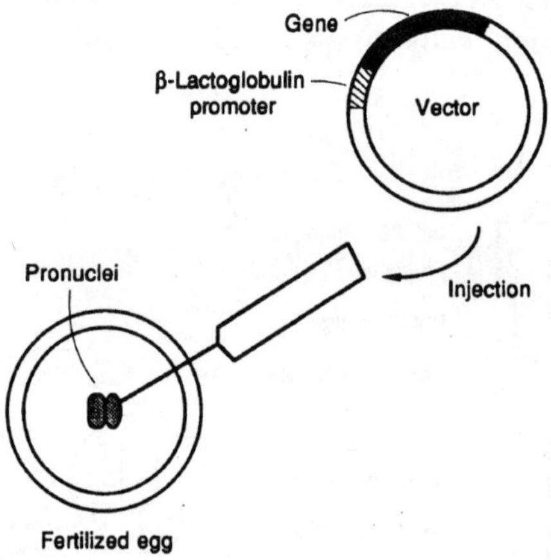

Fig. 10.2. Injecting DNA into pronuclei.

10.7 Nuclear Transfer

The technology of nuclear transfer involves the removal of the nucleus from an unfertilized egg (oocyte) taken from an animal soon after ovulation by using a dedicated needle operated under a high power microscope. The resulting cell, now devoid of genetic materials, is fused with a the donor cell carrying its complete nucleus. The fused cells then develop like a normal embryo, and finally implanted into the uterus of a surrogate mother to produce offspring (Fig. 10.3). Instead of using a whole donor cell to fuse with the recipient cell, the donor cell nucleus can be removed and transferred by injecting the DNA directly into the recipient cell (see Section 22.1).

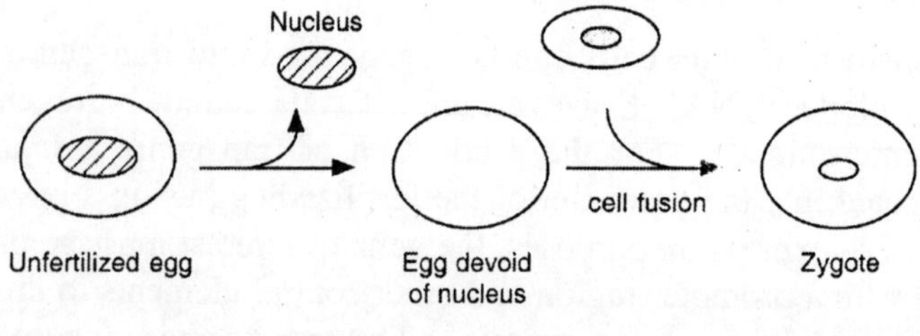

Fig. 10.3. The nuclear transfer process.

Review

1. Which transformation method(s) uses (A) mechanical means of introducing DNA into cells? (B) Biological means? (C) Chemical means?
2. What are the three methods of *Agrobacterium* transformation used for plant cells?
3. Transformation of plant and mammalian cells often results in random insertion of DNA into the cell genome. What are the disadvantages of random insertion?
4. Describe the sequence of steps in performing nuclear transfer.

CHAPTER 11

Isolating Genes for Cloning

Gene cloning requires, as an initial step, isolation of a specific gene encoding for the protein of interest. Locating and selecting a single gene among thousands of genes in a genome is not a simple task.

11.1 The Genomic Library

For prokaryotes, identification of a particular gene is usually made by first constructing a genomic library (Fig. 11.1). The total genomic DNA is isolated, purified, and digested with a restriction enzyme. The short DNA fragments are then cloned into a suitable vector, for example, bacteriophage λ vectors or phagemids. The resulting recombinant λ DNAs are assembled into phage particles by *in vitro* packaging, with all the required phage proteins provided by λ phage mutants that cannot replicate. Now, we have a library of all genomic DNA in short fragments (usually average ~15 kb) cloned into bacteriophage λ vectors packaged into viable phage particles.

The phage particles are used to infect *E. coli* cells. Phage transfection results in clear plaques on a bacterial lawn. Each plaque corresponds to a single phage infection. The next step is to screen the plaques for the clone(s) containing the gene of interest. A commonly used technique is DNA hybridization as described (see Section 8.6). A radiolabeled short DNA probe complementary to the gene sequence is used to identify the particular recombinant clone. This technique obviously implies some prior knowledge of a short segment of the gene sequence for constructing oligonucleotide labels. The sequence (usually 18 to 22 bases) can be deduced from (1) sequences from comparative species, (2) the known sequence of the protein encoded by the gene, or (3) N-terminal or peptide sequencing of the protein.

An alternative is immunological method, based on the detection of the translation product of the gene. This screening method requires that the λ vector used is an expression vector (such as λgt11, see Sections 9.12 and 12.2) the insertion of the gene is in frame, and the availability of antibodies raised against the protein. A third alternative method used for screening libraries involves the use of PCR to amplify the gene of interest. This requires some knowledge of short sequences at the 5' and 3'end to synthesis primers for the reaction. With the PCR, the genomic DNA or cDNA can be used directly without the need for making clones.

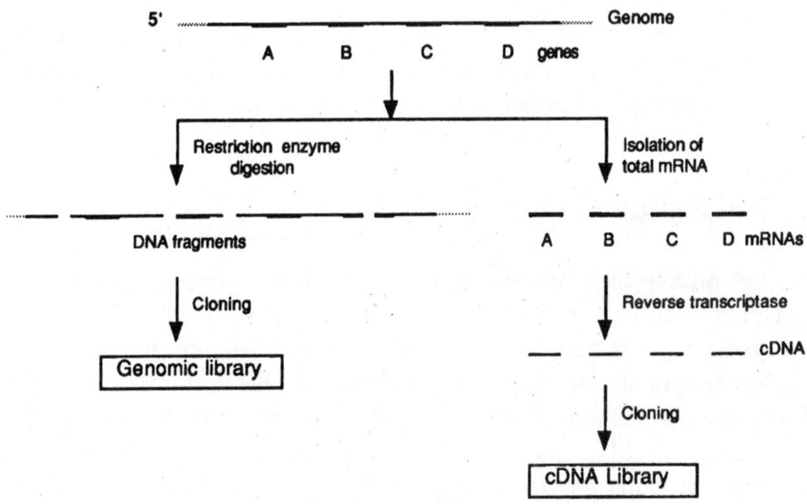

Fig. 11.1. Construction of genomic and cDNA libraries.

11.2 The cDNA Library

Construction of genomic library for gene isolation is usually applied to prokaryotes. For fungi, plants and animals, the identification of a gene from genomic library is not suitable for at least two reasons: (1) The large genome size in plants and animals requires screening of an astronomical numbers of clones for the gene of interest; (2) Eukaryotic genes contain introns, non-coding regions that cause complications in subsequent expression. The use of a cDNA library can circumvent these problems..

The construction of a cDNA library begins with the isolation of total RNAs from a specific cell type that produces the protein of interest, the isolation of mRNAs from the total RNAs, followed by the conversion of the mRNA molecules to complementary DNA (cDNA) strands (Fig. 11.2).

The isolation of mRNA makes use of the fact that eukaryotic mRNA has poly(A) tail at the 3' end (see Section 5.5.3), that can hybridize with oligonucleotides of poly(T) immobilized on a column matrix. When total RNA extract is applied onto such affinity column, mRNA will be retained, while the bulk of the cellular materials pass through. The isolated mRNA sample is then treated with reverse transcriptase to synthesize complementary DNA strands (cDNA) using the mRNAs as templates, forming RNA:cDNA hybrids. The RNA strands in the hybrids are nicked with ribonuclease, and nick-translated by DNA polymerase I to replace the RNA strand with a new (second) cDNA strand, forming dsDNA molecules.

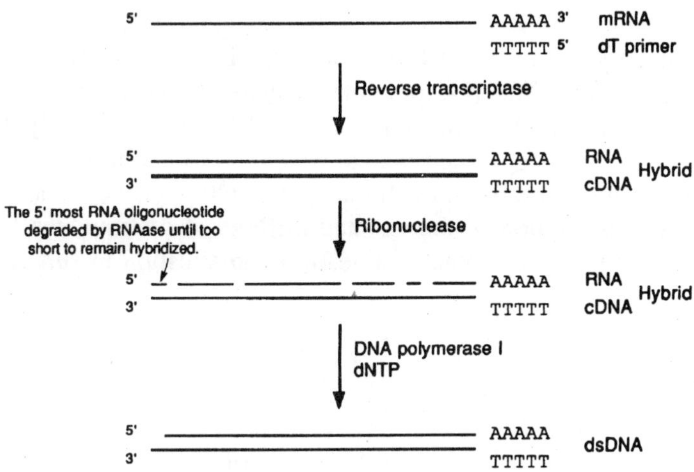

Fig. 11.2. Reverse transcription of RNA to complementary DNA.

The resulting cDNAs are inserted into a suitable vector (plasmids or bacteriophage λvectors), and introduced into *E. coli*. The cDNA library thus constructed is screened for the clones carrying the gene of interesting using DNA hybridization or immunological detection method. The gene of interest can also be isolated by PCR amplification. The identified clone(s) is then cultured and the recombinant DNA is purified, using the cDNA mixture directly without cloning. The gene sequence is excised from the vector, followed by determination of the complete nucleotide sequence. The isolated gene sequence must be determined for at least two reasons: (1) To confirm that the identified sequence is the gene of interest;

(2) For proper construction of the regulatory region and in frame ligation for gene expression. A full-length gene may not be identified in the first trial. The gene isolated may be truncated (missing segments at the 5' end, for example). In such cases, the partial gene can be used as a probe to "fish" for longer length cDNA clones.

11.3 Choosing The Right Cell Types For mRNA Isolation

Choosing the right cell type for isolating mRNA is key for constructing a cDNA library. All cells in an organism contain the some genome composition, but different cell types express different sets of genes. The development and functions specialized in any particular cell type requires only the expression of certain number of genes, with a majority of the genes in the genome silent (not operational). The isolation of bovine β-lactoglobulin (a milk protein) gene, for example, requires the use of bovine mammary tissue for mRNA isolation. Likewise, human pancreatic tissue is the source of mRNA for human insulin gene isolation.

By targeting the right cell type for mRNA isolation, only the active genes will be included in the final construction of the library. The number of clones need to be screened will be significantly reduced. The formation of reverse transcription of the isolated mRNAs also eliminates the intron sequences, which would otherwise lead to complication in subsequent expression.

Review

1. Why is it necessary to construct a cDNA library for isolating eukaryotic genes? What are the advantages over a genomic library?
2. Why is it that genomic libraries are used for isolating genes in bacteria? Why are cDNA libraries not used for prokaryotes?
3. Why are phage λ vectors used in library construction? What other vectors do you think are also suitable for the same purpose?
4. In the construction of cDNA libraries, it is important to isolated mRNA from a specific cell type producing the protein of interest. Explain the underlying reasons.
5. Why is it necessary to determine the sequence of the cDNA after its isolation?
6. List the enzymes used in the construction of cDNA libraries, and describe their functional roles in the procedure.

Part Three

Impact of Gene Cloning

- Applications in Agriculture

CHAPTER 12

Improving Tomato Quality by Antisense RNA

Fruit ripening involves biochemical and physiological changes that are primary factors influencing quality attributes, such as color, flavor and texture of the product. Tissue softening of fruits during ripening is the result of solubilization of the cell wall by a group of enzymes. One of the key enzymes is polygalacturonase (PG), functioning in the breakdown of pectin, a polymer of galacturonic acids that forms part of the structural support in cell wall.

The texture of tomato fruit is a major quality consideration of commercial importance in both fresh market and commercial processing. Tomatoes sold in markets are picked from the field when they are green, stored under low temperature, and gassed with ethylene to trigger fruit coloration and ripening. Recombinant DNA technology is now capable of engineering tomatoes that soften slowly and can be left to ripe on the vine, with full development of color and flavor. The increase in firmness allows the tomatoes to be handled and shipped with minimized damage. The tomatoes also have enhanced rheological characteristics (such as viscosity), making them suitable for various processing applications. These engineered tomatoes are controlled by inhibition of the expression of the PG gene by antisense RNA. The basic idea of this technique is to introduce into the plant an RNA molecule that is complementary to the mRNA of the PG gene.

12.1 Antisense RNA

In normal gene function, the gene is transcribed into mRNA, which is translated into the enzyme PG. If one can introduce a piece of RNA with a sequence complementary to that of the PG mRNA, this piece of RNA would be able to bind to the mRNA preventing the translation of the mRNA and consequently the production of the enzyme. The RNA molecule that is complementary to the mRNA is called antisense RNA, and the mRNA is the sense RNA (Fig. 12.1). In practice, quite often the gene (in this case, the PG gene) is inserted in a reverse orientation (back-

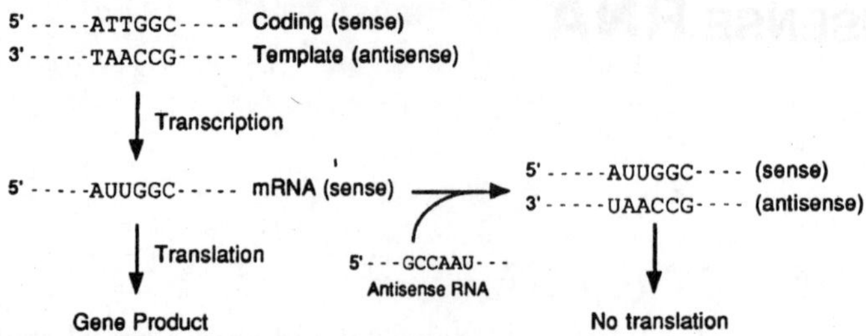

Fig. 12.1. Inhibitory action of antisense RNA.

to-front with respect to the regulatory region) into the vector, so that the coding strand becomes the template for transcription. The mRNA transcript becomes the antisense RNA, and base-pair with the sense RNA (Fig. 12.2). Since the expression of the PG gene is blocked, the plant loses its ability to produce polygalacturonase.

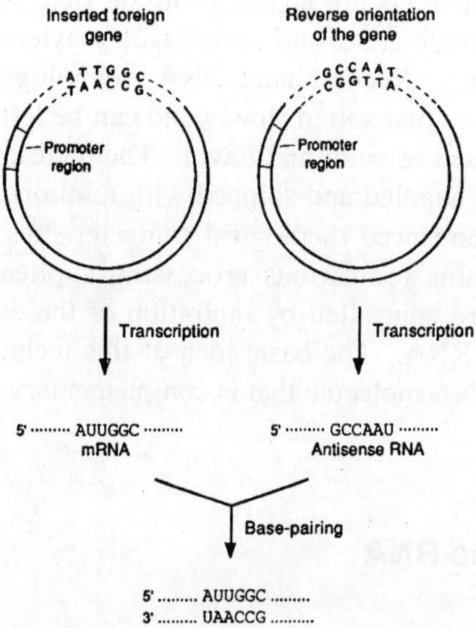

Fig. 12.2. Insertion of a gene in a reverse orientation to produce antisense RNA.

12.2 A Strategy for Engineering Tomatoes with Antisense RNA

The following description outlines one of the cloning strategies used to generate engineered tomatoes with reduced PG activity by antisense RNA (Sheehy et al. 1987. *Mol. Gen. Genet.* 208, 30-36; Sheehy et al. 1988. *Proc. Natl. Acad. Sci.* USA 85, 8805-8809).

(1) Isolation of cDNA of a tomato PG gene. The total RNA was extracted from ripe tomato fruit tissues and mRNA is purified using a poly(T) affinity column. Eukaryotic mRNAs contain poly(A) sequence at the 3' end that can bind to poly(T) on a gel matrix in the column. The poly(A) RNAs obtained was converted to ss cDNAs by reverse transcriptase (which catalyzes the synthesis of DNA from an RNA template) (see Sections 7.3.3 and 11.2). Second strand synthesis produced the ds cDNA (Fig. 12.3).

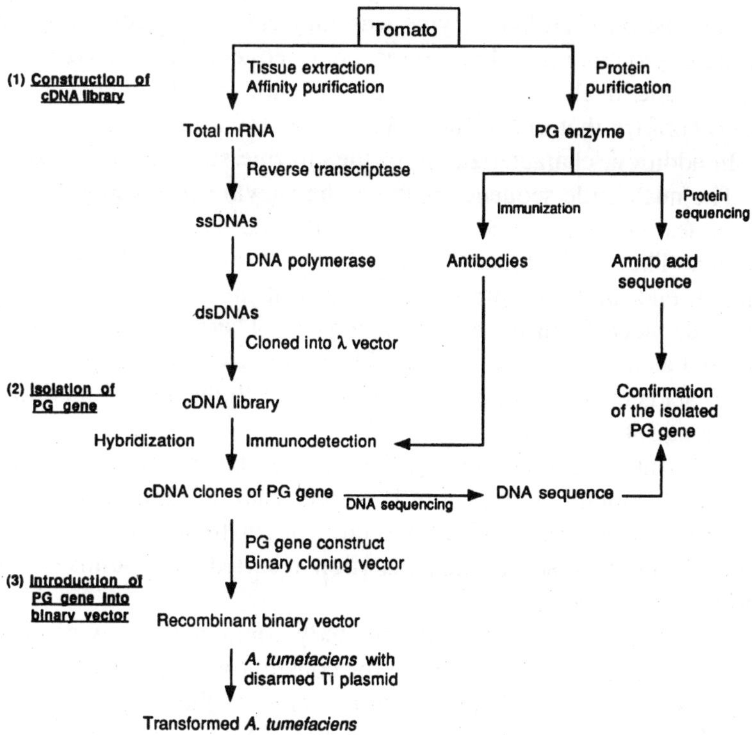

Fig. 12.3. Strategy for engineering tomatoes by antisense RNA.

The cDNA ends were blunted using polymerase I Klenow fragment, followed by ligation with *Eco*RI linkers (see Section 7.1). This step was to create *Eco*RI cohesive ends for the cDNAs so that they could be inserted into the unique *Eco*RI site in λ vectors. The cDNAs (with cohesive *Eco*RI) were cloned into the unique *Eco*RI site of λgt10 and λgt11 vectors, followed by *in vitro* packaging (see Section 9.1.2). Infection of *E. coli* cells by the generated virus particles yielded plaques that were screened by hybridization (for λgt10 which is not an expression vector), or immunological detection (for λgt11 which is an expression vector) (see Sections 8.6 and 8.7). The screening step using immunological detection methods implies that the PG enzyme had to be purified from tomato tissue (by conventional protein purification procedures) for antibody preparation. The transformants (i.e. immuno-positive clones) identified in the screening of λgt11 library were used as the source for DNA probe to screen the λgt10 library by hybridization.

The identified cDNAs from the two libraries were then subcloned into M13 vectors for DNA sequencing (see Sections 8.8 and 11.2). At the same time, the purified PG enzyme was subjected to peptide mapping and amino acid sequencing. The amino acid sequence predicted from the cDNA, was identical to the amino acid sequence of the protein. These results thus confirm that the cDNA isolated from the library was indeed a PG gene. In addition, characterization of the enzyme structure was made possible. The nucleotide sequence predicts the enzyme containing 373 amino acid residues with calculated MW of 40,279, and 4 potential sites for glycosylation. The purified mature protein was 71 amino acids shorter at the N-terminal end, and 13 amino acids shorter at the C-terminus, compared with that deduced from nucleotide sequence. Therefore, the protein was synthesized as a proenzyme with the 71 amino acid signal peptide and the 13 amino acids at the C-terminus cleaved in posttranslational modification (see Sections 3.4 and 6.2).

(2) Introduction of cDNA PG gene into binary cloning vector. A 1.6 kb cDNA containing the entire PG open reading frame was inserted in reverse orientation into a plasmid vector, downstream of the CaMV35S promoter. The reverse-oriented insertion resulted in an antisense transcription of the PG gene.

The binary vector used in the study contained: (1) A cauliflower mosaic virus (CaMV) 35S promoter for constitutive expression of the PG gene; (2) A transcript 7 3' termination region of the Ti plasmid for the control of transcription termination signal; (3) A *neo* gene encoding neomycin phosphotransferase II (NPTII) that confers kanamycin resistance. (This is a dominant selectable marker that serves to screen transformed cells/transgenic plants (See Section 9.3.3.); (4) The left and right border sequences of T-DNA; (5) A pUC plasmid *ori* for replication in a bacterial system (Fig. 12.4) (see Section 9.3.1).

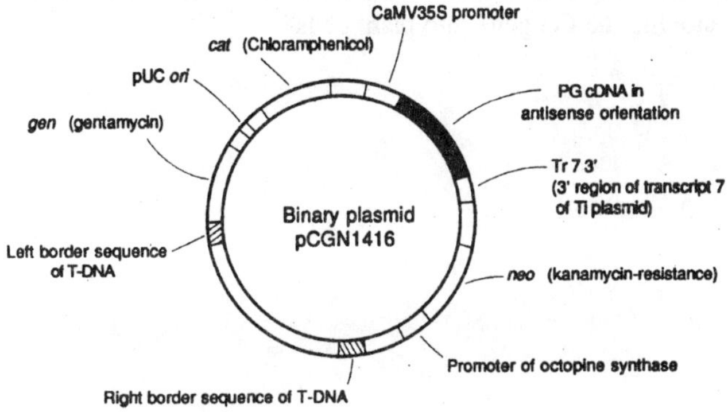

Fig. 12.4. Binary cloning vector used in antisense construction of PG gene.

(3) Introduction of the recombinant vector into tomato plant cells. The recombinant binary vector was introduced into *Agrobacterium tumefaciens* containing disarmed T plasmids. Transformants were selected by the *neo* gene marker, and used to infect tomato cotyledon sections by co-cultivation. Transformed plant cells selected by kanamycin, were regenerated into plants, and screened for NPTII activity in leaf tissues (see Section 10.3).

Review

1. Given the following DNA sequence,

   5′---ACGTGCCTCG---3′ Coding strand
   3′---TGCACGGAGC---5′

   (a) Which strand is the sense strand? Antisense strand?
   (b) What is the sequence of mRNA after transcrition?
   (c) What is the sequence of the antisense RNA?
2. In the example of engineering tomatoes with antisense RNA, λgt10 and λgt11 vectors were used for library construction. What are the differences between these two vectors? Can you suggest other vectors that may also be used for library construction?
3. A binary vector system was used to introduce the gene in tomato tissues. What would be an alternative system? Give a description and show how the other system could be utilized in this case.

4. The PG gene was isolated from a cDNA library in the described cloning strategy. Why was it necessary to use a cDNA library?
5. M13 sequencing was used to obtain the nucleotide sequence of the isolated PG gene. What is an alternative to the use of M13 vectors?
6. What are the unique functions of the CaMV35S promoter that make it desirable for cloning the PG gene into plant cells?

CHAPTER 13

TRANSGENIC CROPS ENGINEERED WITH INSECTICIDAL ACTIVITY

Public concerns over the environmental and health effects of chemical pesticides have intensified the effort to search for alternatives. One of the attractive options is the use of biopesticides from microorganisms. The role of biopesticides in crop protection is not new. In fact the first such product, based on the insecticidal activity of *Bacillus thuringiensis*, has been in commercial applications for more than 20 years. With the advent of recombinant DNA technology, scientists have produced transgenic crop plants engineered with insecticidal activity.

13.1 *Bacillus thuringiensis* Toxins

Bacillus thuringiensis (*Bt*) is a spore-forming bacterium that is specifically lethal to a number of insect pests. Most *Bt* strains contain activity against insect species in the orders Lepidoptera (cotton bollworm, tomato fruitworm), Diptera (mosquito, blackfly), and Coleoptera (e.g. Colorado potato beetle). The insecticidal activity resides in a crystal protein, called δ-endotoxin within the cell. The protein, which is 13 to 14 kD in molecular weight, is released together with the spores when the bacterial cell lyses. When ingested by an insect, the crystal protein is solubilized in the alkaline conditions of the insect's midgut, followed by proteolysis (cleavage) into an active N-terminal segment peptide. This toxic peptide binds to the surface of the cells lining the gut, and penetrates itself into the cell membrane. The host cell is perforated and ruptured due to increased internal pressure.

δ-Endotoxins are classified according to their activities into CryI (active against Lepidoptera), CryII (active against Lepidoptera and Diptera), CryIII (active against Coleoptera), CryIV (active against Diptera), and CryV (active against Lepidoptera and Coleoptera). Each major class is further grouped according to sequence homology. CryI proteins are divided into 6 groups: 1A(a), 1A(b), 1A(c), 1B, 1C, 1D, and so on. The majority of the work on transgenic crops has focused on tobacco, cotton, and tomato plants transformed with *cryI* genes which have activity against lepidopterans. The following example describes one of the strategies used for the construction and generation of insect-resistant cotton plants.

13.2 Cloning of the *cry* Gene into Cotton Plants

In engineering insect resistant cotton plants, two specific modifications were used to improve the expression of *cry* gene (Perlak et al. 1990. *Bio/Technology* 8, 939-943).

13.2.1 Modifying The *cry* Gene

(1) The N-terminal segment of the *cryIA* gene was partially modified by substituting many of the A and T nucleotides to G and C, without changing the amino acid sequence. The resulting GC-rich gene has been shown to enhance the expression level by 10-100 fold.

(2) The cauliflower mosaic virus (CaMV) 35S promoter used for controlling the *cryIA* gene was constructed with duplicated enhancer (transcription activating) sequences (upstream of the TATA box) (see Section 5.5.1). The transcriptional activity was 10 fold higher than that of the natural CaMV35S promoter. A 3' polyadenylation signal sequence derived from the nopaline synthase gene was placed downstream of the *cryIA* gene. (The 3' polyA sequence serves as a termination signal for transcription.)

13.2.2 The Intermediate Vector

The intermediate vector contains the following elements (Fig. 13.1):

1. Selectable markers: A *neo* gene that confers resistance to kanamycin, and a mutated 5-endopyruvylshimate 3-phosphate synthase (EPSPS) gene that confers resistance to the herbicide glyphosate (see Section 14.1).
2. A Ti plasmid homology sequence to facilitate recombination.

3 ColE1 *ori* from pBR322.
4. A right border sequence of T-DNA (see Section 9.3.2).

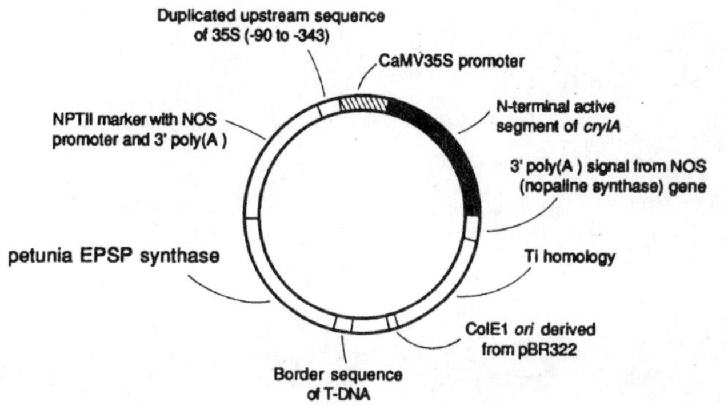

Fig. 13.1. Intermediate vector for cloning the *cryI*A gene

13.2.3 Transformation by *Agrobacterium*

The vector was used to transform cotton by inoculating cotyledon explants with *Agrobacterium* containing cointegrates of a disarmed plasmid and the intermediate vector (see Section 10.3). The explants transformed with *Agrobacterium* carrying the intermediate vector, were kanamycin resistant, and the plantlets produced from kanamycin-selected cotyledons survived in the presence of glyphosate. The transgenic mature plants were screened for *Bt* protein by an immunological detection method (ELISA). Positive plants were grown and cotton balls were artificially infested with bollworm eggs. Approximately 70-75% of the balls survived the infestation.

Review

1. What are the rationales for introducing Bt toxin genes into crop plants?
2. Describe the two specific modifications used to improve the expression of the cry gene.
3. In the example given, what type of vector system was used?
4. What is the purpose of incorporating a mutant EPSPS gene into the vector?
5. What are the main features in an intermediate vector? Describe their functions, and how they participate in the formation of a cointegrative vector?

CHAPTER 14

TRANSGENIC CROPS CONFERRED WITH HERBICIDE RESISTANCE

Herbicides act by inhibiting the function of a protein or an enzyme that is involved in certain vital biological possesses. For example, glyphosate and chlorosulfuron inactivate key enzymes in the biosynthesis of amino acids. Bromoxynil and atrazine interfere with photosynthesis by binding with the Q_B protein. Herbicides are therefore non-selective, because the biosynthetic pathways involved are present in both the weed and the crop plant. The effectiveness of a herbicide on the control of weeds depends on the differential uptake or metabolism of the herbicide between the weed and the crop.

Herbicide	Physiological Effect	Specific Target	Modifications
Glyphosate [*N*–(phosphonomethyl)glycine] 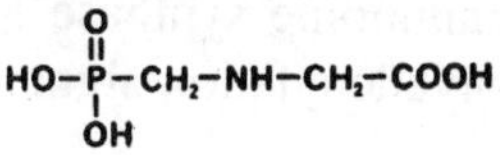	Inhibits aromatic amino acid synthesis.	Inhibits 5–enolpyruvyl shimate–3–phosphate	1. Cloning of a mutant EPSPS gene. 2. Amplification to increase the concentration of EPSPS.
Chlorosulfuron [sulfonylurea] 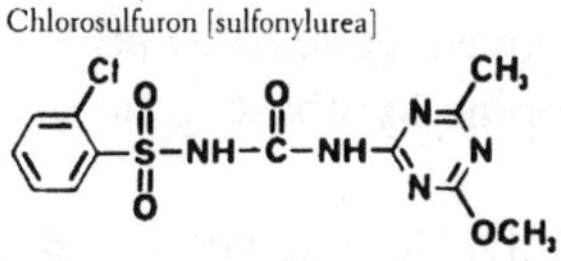	Inhibits branched-chain amino acid synthesis.	Inhibits acetolactate synthase (ALS).	Cloning of a mutant ALS insensitive to inhibition.
Bromoxynil 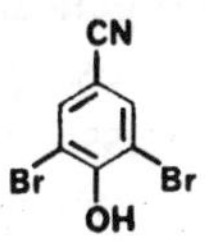	Inhibits photosynthetic electron transport	Binding to Q_B protein.	Introduction of bacterial *bxn* gene (nitrilase) into plants to detoxify bromoxynil.
Atrazine [triazine] 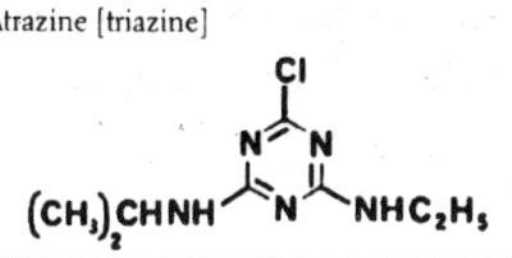	Inhibits photosynthetic electron transport.	Binding to Q_B protein.	1. Introdution of mutant *psbaA* gene to produce nonbinding Q protein. 2. Introduction of glutathione–S–transferase to detoxify the herbicide.

Genetic engineering has provided a means to confer herbicide resistance in plants. The fact that many common herbicides act on a single target renders the introduction of herbicide tolerance into crops an achievable alternative. Various approaches are available but all involve the transfer of a single gene into plants.

14.1 Glyphosate

Glyphosate is one of the widely used non-selective herbicides. It inhibits 5-enolpyruvylshikimate-3-phosphate synthase (EPSPS), a key enzyme in the synthetic pathway for aromatic amino acids. Two different approaches have been taken to produce glyphosate resistance plants. One method involves the overproduction of EPSPS by transferring the gene under the CaMV35S promoter. Alternatively, a mutant gene encoding EPSPS that is insensitive to glyphosate is used. The *aroA* gene selected from a glyphosate resistant mutant strain of *Salmonella typhimurium* has been isolated which encodes for a mutant form of EPSPS resistant to inhibition by glyphosate. The *aroA* gene has been cloned into several plants.

14.2 Cloning of the *aroA* gene

In the transfer of the *aroA* gene into tomato (Fillatti et al. 1987. *Bio/Technology* 5, 726-730.), a binary cloning vector was used in the transformation system (Fig. 14.1).

(1) The *aroA* gene was regulated by the mannopine synthase gene (*mas*) promoter, and contained at the 3' end a polyadenylation signal sequence derived from the large tumor gene (*tml*).

(2) A *neo* gene was fused to the octopin synthase gene (*ocs*) promoter and a second *neo* gene was fused to the mannopine synthase gene (*mas*) promoter.

(3) The left and right border sequences of T-DNA. The *neo* and *aroA* genes are prokaryotic in origin. It is necessary to put the genes under the control of eukaryotic transcriptional regulatory sequences (promoters, polyadenylation signal sequences) to ensure efficient expression. The *ocs* and *mas* genes are genes originated from the T-DNA of T-plasmids. They have been widely used in the construction of plant vectors because the promoter elements and polyadenylation signals of these genes are eukaryotic in character.

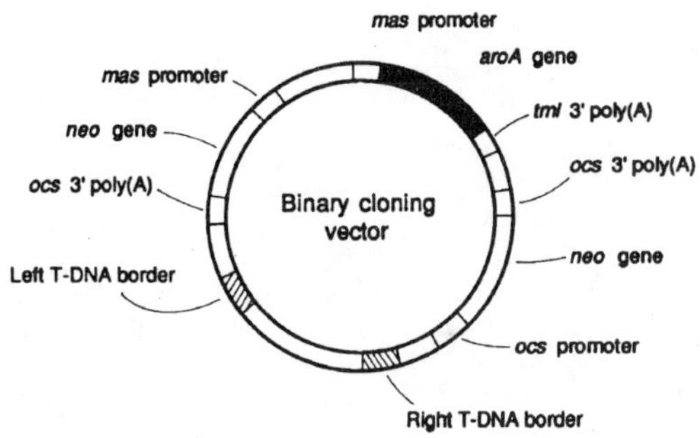

Fig. 14.1. Binary cloning vector used for cloning the *aroA* gene.

The binary cloning vector was introduced into *Agrobacterium fumefaciens* strain that carried a disarmed Ti plasmid (with the T-DNA deleted, but the *vir* region retained.) The disarmed Ti plasmid acts as a helper plasmid to mediate the transfer of T-DNA region in the binary cloning vector to be integrated into the plant chromosomal DNA carrying with it the *aroA* gene. Transformed tomato cells would develop resistance to kanamycin present in the growth medium. The production of the EPSPS enzyme encoded by the *aroA* gene was demonstrated by Western blot analysis of leave tissues. The presence of *aroA* gene in tomato leaves was confirmed by Southern blot.

Review

1. What is the gene product of the *aroA* gene? Why was it chosen for engineering herbicide resistance in plants?
2.

Herbicide	Mechanism of action
Arazine	
Bromoxynil	
Chlorosulfuron	
Glyphosate	

3. In the cloning of the *aroA* gene, the binary cloning vector was introduced into *Agrobacterium* carrying a disarmed Ti plasmid. Why was the Ti plasmid "disarmed"?
4. The left and right border sequences of T-DNA were included in the binary cloning vector (Fig. 13.1). Explain the reason why this was done.

CHAPTER 15

GROWTH ENHANCEMENT IN TRANSGENIC FISH

The aquaculture industry in the U.S. produces ~500 million pounds of processed fish yearly. Much of the improvement in fish farming has been done by traditional breeding methods. In the past two decades, there has been marked progress in employing recombinant DNA technology to produce transgenic fish with desirable traits, such as increased growth rate and disease resistance.

Fish is particularly suitable for transgenic manipulation, because most fish species reproduce by external fertilization. Fish eggs and sperms can be collected in large numbers by relatively simple operations, fertilization is facilitated by gentle mixing and stirring, and fertilized eggs can be reared under controlled conditions.

15.1 Gene Transfer in Fish

Gene transfer is usually done by direct microinjection of a gene construct into the cytoplasm of a fertilized egg within 1-4 hours after fertilization. This procedure is in contrast to the injection into the pronuclei of a fertilized egg in transgenic mice, cows, and other animals (see Sections 10.6 and 21.1). The reason is that the pronuclei of fertilized fish eggs are difficult to locate due to the presence of a large yolk mass.

Fertilized fish eggs contain a tough shell (chorion) that makes it difficult for micro-needles to penetrate. A common practice is to remove the chorion by enzyme digestion, or cut open a small hole by microsurgery before injection. Salmon eggs are large (5-6 mm in diameter), the micropile, an opening for sperm penetration during fertilization, is visible, and can be used for needle insertion. An alternative is electroporation, which have been shown to be effective in transferring genes to fertilized eggs of carp and catfish. Electroporation offers the advantage of treating a large number of fertilized eggs in one batch, in contrast to the handling of individual eggs separately in microinjection (see Section 10.2).

The success rate of DNA integration into fish genome ranges from 10 to 75%. In most cases, multiple copies of gene are inserted in tandem at random locations in the genome. The level and specificity of gene expression depends on the promoter/enhancer utilized in the gene construct. The majority of transgenic fish studies use promoters from animal viruses such as simian virus 40 (SV10) and rous sarcoma virus (RSV). Fish gene promoters have increasingly been used. In the early studies, mammalian genes, such as human growth hormone, were used. Some reports describe the use of genes from related fish species. Successful gene transfer must be evaluated by monitoring the expression of the transferred gene in the offspring of parental transgenic fish. The characteristic phenotype should be observed in the transgenic fish.

15.2 Cloning Salmons with a Chimeric Growth Hormone Gene

An "all fish" chimeric growth hormone gene construct was developed for gene transfer in Atlantic salmon for enhancement of somatic growth rate (Du et al. 1992. *Bio/Technology* 10, 176-180). The all fish gene construct contained a chinook salmon growth hormone (GH) cDNA coding region with the 5' and 3' untranslated regions flanked by a promoter and a 3' polyA sequence of an antifreeze protein gene isolated from ocean pout (Fig. 15.1). The use of an antifreeze protein gene promoter ensures that the expression of the GH gene is functional in the tissues. (Antifreeze proteins act to protect several fish species in subzero sea water temperatures by inhibiting ice crystal formation in blood plasma.)

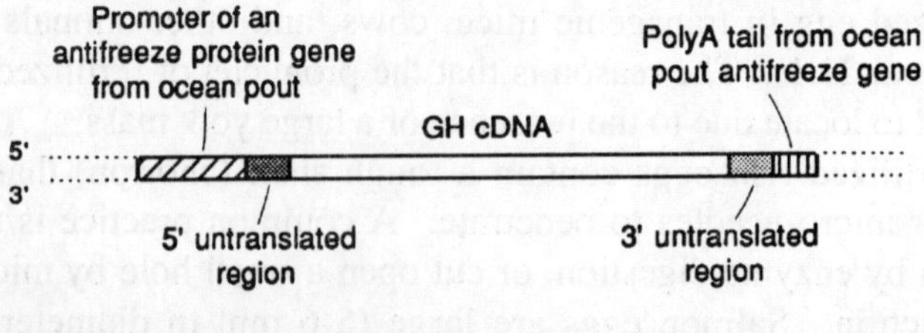

Fig. 15.1. Chimeric growth hormone gene for cloning salmon.

Each salmon egg was microinjected through the micropyle with 3-5 μl DNA (~106 copies) of the GH gene construct following by incubation until hatch. The fish were screened for the presence of the GH gene construct at the 11 and 14 months by PCR analysis of DNA isolated from red blood cells. Growth rates of both transgenic and non-transgenic salmon were monitored. At one year old, a 2- to 6-fold enhancement of growth was observed. It should be noted that the insertion of a GH gene is to *enhance* the growth rate by increasing the amount of growth hormone that is endogenous to the fish, since fish contains GH gene of its own.

Review

1. What are the advantages of using an antifreeze protein gene promoter in cloning the growth hormone gene into fish?
2. The salmon growth hormone (GH) used in the study was obtained from screening a cDNA library. What was it necessary to use a cDNA library?
3. How is a cDNA library constructed? How is it different from a genomic library? Under what circumstances must a cDNA library be used for gene isolation?

Part Four

Impact of Gene Cloning

- Applications in Medicine

CHAPTER 16

MICROBIAL PRODUCTION OF RECOMBINANT HUMAN INSULIN

The early success of recombinant DNA technology relies heavily on the elucidation of the biological possesses at the molecular level in microbial systems. The first commercial application is realized in the microbial production of human insulin.

16.1 Structure and Action of Insulin

The primary role of insulin is to control the absorption of glucose from the bloodstream into cells where glucose is utilized as an energy source or converted into glycogen for storage. Insulin functions to regulate the level of glucose in blood. Carbohydrates, such as starch, taken in the diet are digested into glucose, which is transferred to the blood stream. The high level of blood glucose stimulates the pancreatic β cells to release insulin directly into blood stream. Insulin binds to insulin receptors on the surface of a cell, generating signals for movements of glucose transporters to the cell membrane. The glucose transporters aggregate into helical structures creating channels for entrance of glucose molecules into the cells.

Insulin is produced in pancreatic cells as prepro-insulin which contains 4 segments: (1) an N-terminal signal sequence of 16 amino acids, (2) a B chain of 30 amino acids, (3) a C peptide of 33 amino acids, and (4) an A chain of 21 amino acids. In a later stage of the process, the N-terminal and the C peptides are cleaved. Active mature insulin consists of A- and B-chains held together by disulfide bonds (Fig. 16.1).

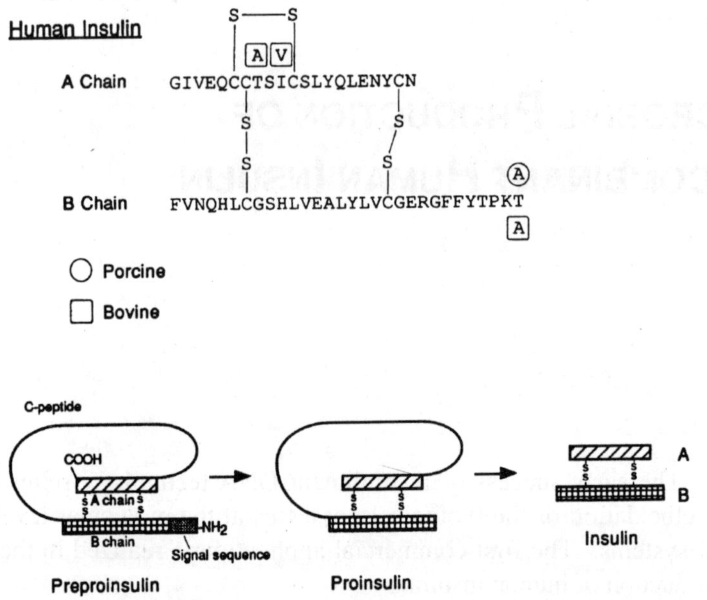

Fig. 16.1. Structure of human insulin and its posttranslational modification.

16.2 Cloning Human Insulin Gene

Before the advent of biotechnology, the insulin used for the treatment of type I (insulin-dependent) diabetes mellitus was obtained from extracting the hormone from porcine or bovine pancreatic tissues. In the early eighties, human insulin produced by recombinant technology entered the pharmaceutical market.

In one of the approaches (Geoddel et al. 1979. *Proc. Natl. Acad. Sci. USA* 76, 106-110), the sequences for the A and B chains were synthesized chemically and inserted separately downstream of the β-galactosidase structural gene controlled by the λ *lac* promoter. The construction was such that the insulin chains would be made as fusion proteins joined by a methionine to the end of the β-galactosidase protein (see Section 9.1.1). The expression vector also contained an Amp^R marker. Transformants were screened by plating on a culture medium containing X-gal and ampicillin. Insulin A chain and B chain transformants were grown to harvest the cells in large quantity. The cells were lysed and the insulin A chain and B chain were purified separately. Because the insulin A gene was fused to the β-galactosidase gene, therefore the insulin protein produced was a β-galactosidase-insulin hybrid.

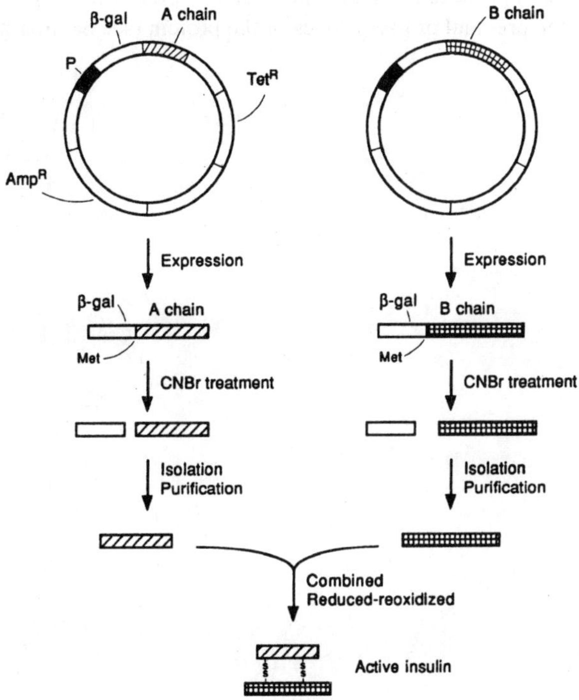

Fig. 16.2. Strategy for cloning and production of human insulin.

This hybrid protein was treated with cyanogen bromide to cleave off the insulin chain at the methionine. Likewise, insulin B chain also underwent the same treatment. The purified insulin A and B chains were mixed and subjected to reduction-reoxidation to insure correct joining of the disulfide bonds (Fig. 16.1). An alternative procedure involves the cloning of human pro-insulin (A-C-B sequence) into bacterial cells. The C-peptide of the expressed protein was enzymatically cleaved to yield the active A-B form.

Review

1. Describe briefly the biological fucntion and mechanism of insulin.
2. In the given example, the human insulin gene was fused to the β-galactosidase gene to produce a β-galactosidase-insulin hybrid. Explain why this was done.

3. Transformed colonies were screened for the human insulin gene by plating the colonies on a medium with X-gal and ampicillin. What color colonies would you pick, blue or white? Explain yours answer.
4. Human insulin is produced in pancreatic cells as prepro-insulin. What are the functions of the pre- and pro-sequences of the protein (see Section 6.2)?

CHAPTER 17

FINDING DISEASE-CAUSING GENES

Among the ~4000 known human genetic disorders, only a handful of disease-causing genes have been mapped. To locate a gene (say an average of 10,000 bp length) in the midst of a 3.2 billion bp chromosomal DNA is hardly a simple task.

In the case that a protein is known to be involved in the genetic disease, the procedure becomes relatively straightforward. The approach is to purify the protein, determine its amino acid sequence, and deduce the gene sequence that encodes the protein. One can then synthesize a probe based on the deduced nucleotide sequence to isolate the gene from the appropriate gene library. Subsequently, the sequence of the target gene isolated from individuals with the genetic disease is compared with that from normal individuals. A mutation in the gene suggests it related to the disease.

17.1 Genetic Linkage

In many incidences, the causative mechanism of the disorder is unknown. Nonetheless, it is still possible to search and identify the gene that is responsible for the disease, using "reverse genetics" - cloning of a gene by pinpointing it to a specific location in a chromosome. The strategy is to look for "markers" that are located near the gene in the chromosome. In the course of meiosis (process of cell division in the production of germ cells - sperms and eggs; see Section 1.5), the homologous chromosomes undergo exchange, a process called recombination. Each chromosome in a germ cell is a genetic combination from the homologous chromosomes in the parental cells. Suppose that two genetic sites (loci), A and B, are located close to one another in the same chromosome. The chance of the two loci staying together as DNA is exchanged during meiosis is high. In other words, it is likely that both will be inherited in future generations. In this case, A and B are said to be "linked". If the two loci are far apart in the same chromosome, the chance of being separated during the recombination process increases (Fig. 17.1).

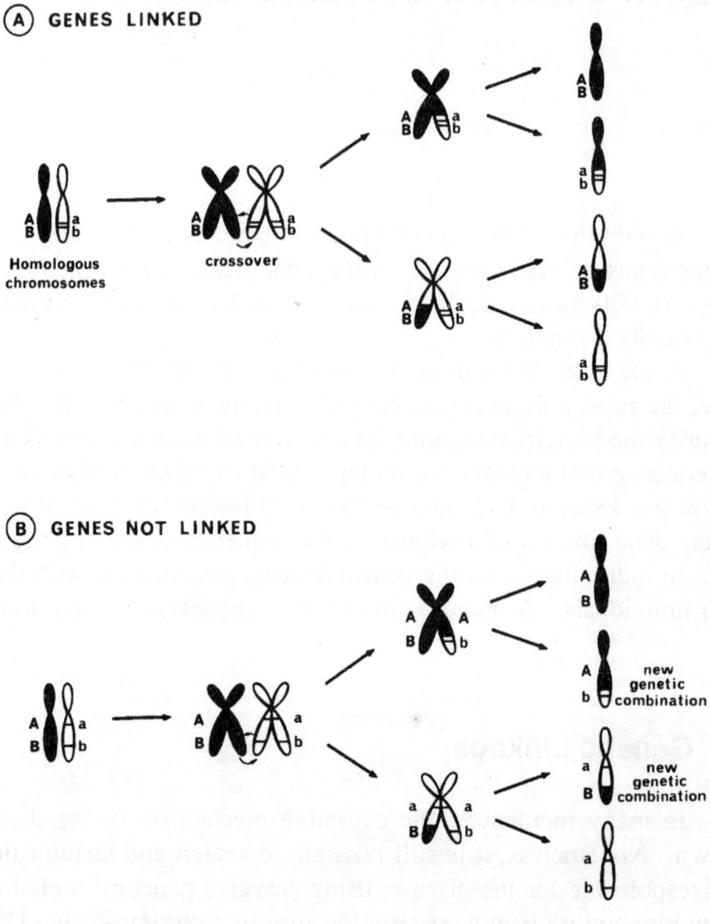

Fig. 17.1. Gene linkage and recombination.

17.1.1 Frequency of Recombination.

A measure of the distance between A and B can be correlated by the frequency of recombination (i.e. How often A and B are separated during the exchange process). A 1% recombination frequency is equivalent to 1 million bp apart, a 5% frequency means they are 5 millions bp, and 0% means complete linkage of the two genes.

There is a limitation in linkage analysis. For a linkage to be detected, A and B must both be heterozygous, existing in normal and mutant forms, e.g. AB/ab (AB in one chromosome, and ab representing mutants of AB, respectively, in the other homolog). If both genes are homozygous, e.g. AB/AB, Ab/Ab, ab/ab, or aB/aB, the recombination even occurred, cannot be detected in the offspring.

17.1.2. Genetic Markers

By analyzing the recombination frequencies among a number of genes, a genetic map can be constructed where the distances of the two loci can be estimated at their relative positions in the chromosome. These genetic loci can then serve as reference points for other new genes. These are known as genetic markers in a genetic map. The majority of markers used in mapping are comprised of polymorphic DNA sequences other than gene sequences. These sequences are variable number tandem repeats (VNTR) possessing two distinct advantages. These tandem repeats exist in many forms (i.e. polymorphic) owing to the variations in the number of repeats and the length of the repeats. This translates to the fact that a given individual will carry different versions of a particular VNTR in a homologous chromosome pair. Moreover, they are readily detectable by restriction fragment length polymorphism analysis (RFLP) using the standard protocol involving restriction digestion, gel electrophoresis, southern blot, and hybridization (see Sections 20.1 and 20.2).

In studying genetic linkage in humans, the first step is to collect blood samples from many patients with the genetic disease and their families. The idea is to look for markers that are always inherited together with the disease, since the closer a marker is to the disease-causing gene, the most likely it is that both will be inherited, (i.e. the recombination frequency is very low). The specific marker thus identified serves as the point of reference for searching the disease-causing gene. In animal studies, crosses between genetically defined parents can be conducted to generate a large number of offspring, and the genetic linkage can be analyzed.

17.2 Positional Cloning

Once the genetic linkage is established between a marker and a gene, then the search for the gene begins at the marker site. Various strategies have been used for this purpose. These include chromosome walking and jumping, and the use of yeast artificial chromosome.

17.2.1 Chromosome Walking.

The chromosome where the marker locates is restriction digested, and used to generate a genomic library of overlapping DNA fragments. The DNA fragment containing the marker is isolated, and the end sequence of the fragment is used to probe for the next overlapping fragment in the chromosome. The end sequence of this second DNA fragment is in turn used for obtaining a third overlapping DNA fragment, extending further along the chromosome. This technique is known as chromosome walking (Fig. 17.2). Each walking step is ~30-40 kb long, which is a rather slow process considering that a recombinant frequency of 1% (considered to be tight linkage) between a gene and a marker is actually equivalent to 1 million bp apart. It is obvious from the above discussion that the primary limitation of search is the size of the DNA fragments that could be cloned (~40 kb in a cosmid vector).

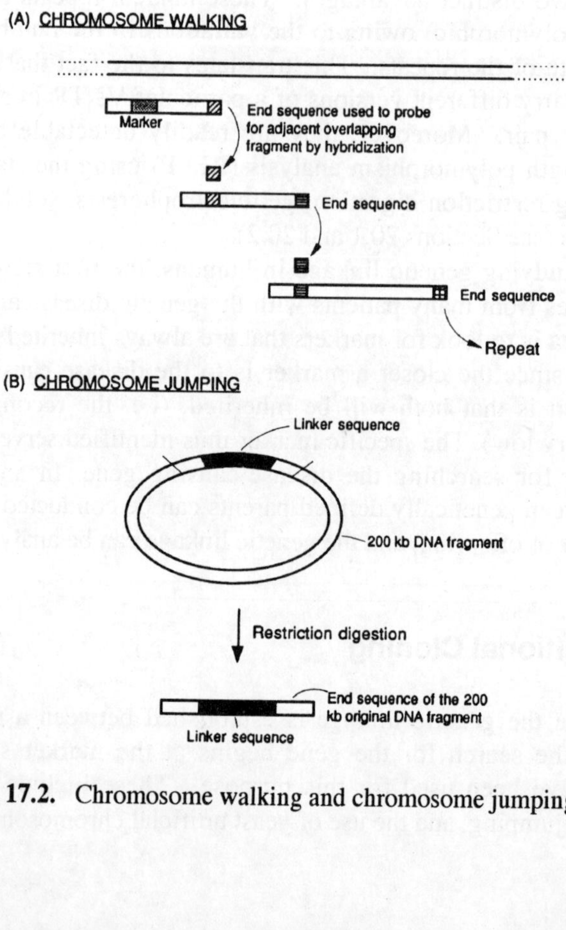

Fig. 17.2. Chromosome walking and chromosome jumping.

17.2.2 Chromosome Jumping.

A technique to circumvent the problem, known as chromosome jumping, enables researchers to jump distances of an average 200 kb, and to resume the research from the end point of each jump. In the procedure, genomic DNA is digested at rare restriction sites, and DNA fragments of ~200 kb in size are isolated. These large DNA fragments are circularized by ligation to a short "tag" (linker) sequence carrying a *E.coli* suppressor tRNA gene. The circularized DNA fragments are digested at many points by a common restriction enzyme (e.g. *Eco*RI) to yield a short fragment (20 kb) consisting of the tag flanked by the two end sequences from the original large DNA fragment (Fig. 17.2). Therefore, each walking step of ~20 kb in this case will correspond to a jump from one end to the other end of a 200 kb fragment.

17.2.3 Yeast Artificial Chromosome.

Another popular alternative is the use of yeast artificial chromosome (YAC) that is capable of cloning DNA inserts in the several hundred kb range (Fig. 17.3).

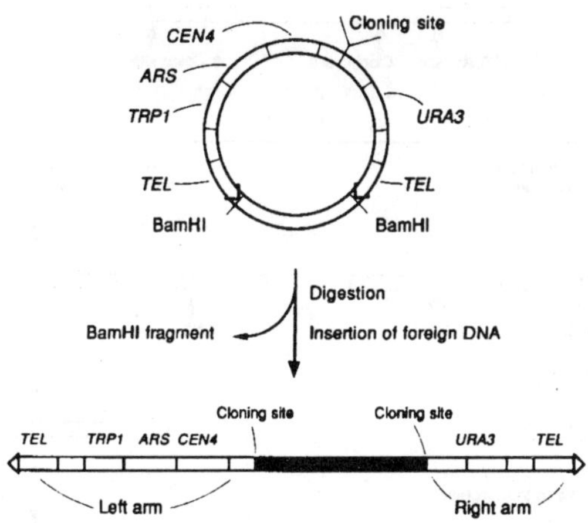

Fig. 17.3. Strategy for using yeast artificial chromosome.

A typical YAC consists of a number of essential yeast chromosomal elements and other structural features: (1) A bacterial origin of replication and antibiotic selectable marker for replication and selection in bacteria; (2) A yeast centromere (*CEN4*) that enables the distribution of the chromosome to daughter cells during cell division; (3) A yeast autonomous replication sequence (*ARS*) for replication in yeast; (4) Two telomeres (*TEL*), end sequences of a chromosome to ensure correct replication; (5) Yeast *URA3* and *TRP1* genes as selective markers for the selection of YAC transformants. The insertion of a large DNA fragment is flanked between the left and right arms of the vector. The left arm consists of *TEL*, *TRP1*, *ARS* and *CEN4*, and the right arm *TEL* and *URA3*. The recombinant vector can be maintained as a linear chromosome in yeast, as a yeast artificial chromosome.

17.3 Exon Amplification

The search along a chromosome continues with frequent testing for coding regions (exons). (The majority of eukaryotic sequence contains introns; 97% of the genome encodes no proteins.) Putative coding sequences can be obtained by exon amplification (also known as exon trapping), a technique based on RNA splicing (Fig. 17.4).

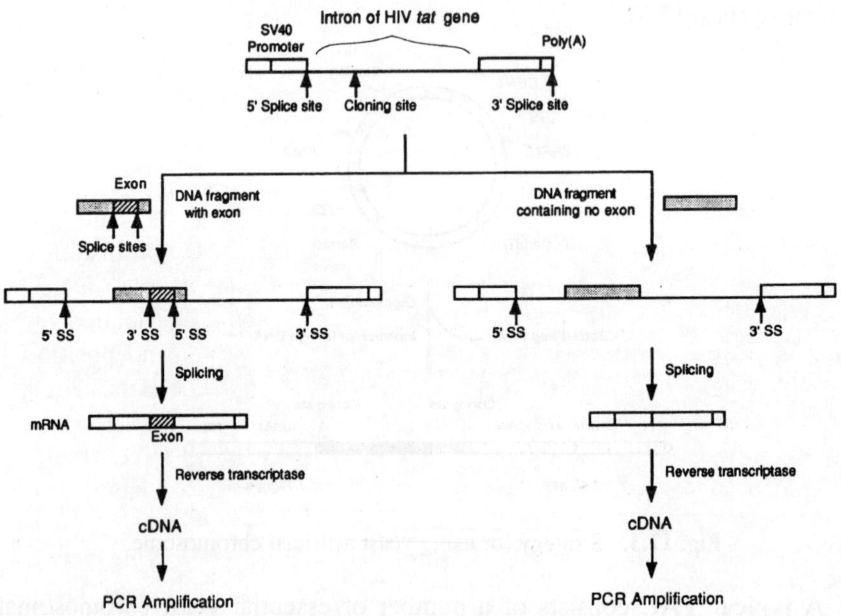

Fig. 17.4. Schematic representation of exon amplification for identifying coding sequences.

Genomic DNA fragments are inserted into an intron segment from the human immunodeficiency virus (HIV-1) *tat* gene flanked by 5' and 3' splice sites. The recombinant DNA constructs are used for cell transfection. If the DNA fragment contains an exon with flanking intron sequence, then the splice sites at the exon-intron junctions will pair with the splice sites of the flanking *tat* intron. In *in vivo* transcription, the mRNA after splicing should acquire the exon derived from the inserted DNA, which can then be amplified by PCR. If the inserted DNA fragment contains no exon, the mRNA will be shorter. The two types of PCR products can be distinguished by size separation on gel electrophoresis.

The exon is excised, sequenced, and checked for the presence of an open reading frame, and methylated GC islands (indicative of transcriptional regulatory sequence). It can also be used to search similar known sequences in other species, based on the assumption that a highly conserved sequence may suggest a coding region of an important gene. The isolated exon sequences can be screened by northern blot for the presence of corresponding RNA from the disease-afflicted tissues. Finally, definitive proof relies on sequence comparison between the putative gene from individuals having the disease with the gene from normal individuals. The causation effect relationship of the gene product to the disease must be established.

17.4 Isolation of The Mouse *Obese* Gene

Obesity is one of the most common causes of serious health problems because it is often associated with type II diabetes (non-insulin dependent), hypertension, and hyperlipidemia. The mouse obsese (*ob*) gene which regulates energy metabolism, has been located and isolated from adipose tissues using linkage analysis, genetic mapping, and positional cloning (Zhang et al. 1994. *Nature* 372, 425-432.). The OB protein encoded by the normal gene acts on the central nervous system to effect a reduction of food intake and increase energy expenditure in mice, resulting in a balanced control of body fat tissues. Mice that are obese have a genotype of *ob/ob*. Both copies of the gene are mutants.

Genetic linkage analysis established the *ob* gene lying between markers *D6Rck13* and *Pax4* on mouse chromosome 6. These two flanking markers were used to probe for clones corresponding to the adjacent regions from the YAC library. Both ends of each YAC were recovered by PCR methods. The ends were sequenced and used to isolate new YACs. The YAC contig (sets of overlapping clones or sequences) was screened for coding regions by exo-trapping. One of the trapped exons was hybridized to northern blot of mouse white adipose tissue, but not of any other tissues. This suggests that the sequence transcribed at significant level in the adipose tissue. The coding sequence also hybridized to vertebrate DNAs in southern blot, demonstrating that the sequence is highly conserved among species. The gene encodes a 4.5 kb mRNA, with an open reading frame of 167 amino acids. The protein was produced by recombinant DNA method and injected daily into obese mice (Halaas et al. 1995. *Science* 269, 543-546). A reduction of 30% of the body weight was observed after a 2-week treatment. Administration of the protein to normal mice resulted in a moderate 12% weight loss.

Review

1. What is "reverse genetic"?
2. How is recombinant frequency related to gene linkage?
3. Why are polymorphic DNA sequences used as (A) genetic markers in linkage analysis, and (2) DNA fing..printing (see Chapter 20)?
4. What is chromosome walking? What is the limitation of this method? Describe how chromosome jumping can overcome some of this limitation.
5. How is exon amplification used to test for putative gene coding sequences in a chromosome?
6. In the isolation of the mouse obese gene, YAC library was constructed. What are the advantages of using YAC? Describe the functions of *ARS*, *CEN4*, *TEL*, *URA3*, and *TRP1*.

CHAPTER 18

HUMAN GENE THERAPY

There are more than 4,000 known inherited disorders. The majority of them have minimal effects, but a few causes physical and mental abnormalities that may be life threatening. Genetic diseases that are candidates for gene therapies include severe combined immunodeficiency, thalassaemia, and cystic fibrosis. Since these genetic diseases are each caused by a single defective gene, one potential treatment is to introduce a normal functional copy of the appropriate gene into the cell tissue that is affected. In effect, the normal (therapeutic) gene augments the defective gene in the patient. Gene therapy is not restricted to only treating genetic disorders. The general technology of transferring genetic materials into a patient is also applied to diseases such as cancer, AIDS, and cardiovascular diseases. Many of the approved clinical trials on gene therapy are for the treatment of diseases other than genetic disorders.

Examples of Known Genetic Disorders:

Genetic Disorder	Incidence	Mutated Gene	Target Cells
Severe combined immunodeficiency	Rare	Adenosine deaminase	Bone marrow or T lymphocyte
Familial hypercholesterolemia	1 in 500	Low-density lipoprotein receptor	Liver
Hemoglobinopathies (Thalassaemia)	1 in 600	α- and β-globin	Bone marrow
Cystic fibrosis	1 in 2,000	Cystic fibrosis T receptor	Lung
Inherited emphysema	1 in 3,500	α-1-antitrypsin	Lung
Duchenne's muscular dystrophy	1 in 10,000	Dystrophin	Muscle
Hemophilia A Hemophilia B	1 in 10,000 1 in 30,000	Factor VIII Factor IX	Liver or fibroblasts

18.1 Physical and Chemical Methods

The techniques of gene transfer can be grouped into (1) physical/chemical methods, and (2) biological (viral) methods. The former methods use a lipid carrier to facilitate the transfer of DNA across a cell membrane. Lipid carriers form complexes with DNA by electrostatic interaction (Fig. 18.1).

Amphipathic lipids which carries both polar groups and a hydrophilic tail in the molecule, can organize into vesicles, forming a liposome structure with the DNA enclosed in the interior. These lipid/DNA complexes (also known as lipoplexes) can be injected directly into cell tissues (for example, tumor tissues in cancer treatment)

Alternatively, DNA can be chemically linked to a ligand that binds to specific receptors on the membrane surface. The cell picks up the DNA-ligand conjugate by binding the ligand to the receptor, and transferring it across the membrane. In most cases, the DNA-ligand conjugate has to be treated further to ensure that the DNA will not be degraded in the lysosome once inside the cell. The final formulation of DNA-ligand conjugate is injected into the blood stream, and circulated to the target tissue.

Various polymers, natural or synthetic, are used to interact and condense the DNA to facilitate the delivery of DNA particles into the cell nucleus. Some studies also explore injecting so-called naked DNA (without a lipid wrap) into the patient (see Section 18.4).

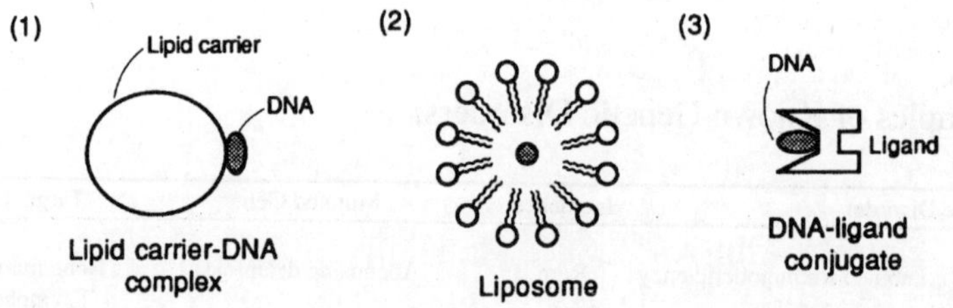

Fig. 18.1. Physical and chemical methods of transferring a therapeutic gene.

For some diseases, *in situ* treatment may be particularly attractive. For example, lipid carriers can be used to inject a gene into a tumor to turn cancerous cells into suicidal cells. In the case of cystic fibrosis that affects the lung, functional copies of the CF transmembrane regulator gene can be introduced directly into the cells lining the respiratory tract. The major disadvantage of all these physical methods is that the effects are transient, and continuous treatment is necessary to insure sustained expression of foreign genes in the tissue cell (See Section 18.4).

18.2 Biological Methods

The biological methods give more stable integration, and comprise majority of the approved clinical trials. Many of these involve the use of viral DNA adapted as vectors (see Section 9.4.4). The most advanced are retrovirus, adenovirus, and adeno-associated virus. A majority of gene-therapy clinical protocols in North America and Europe involve the use of viral vectors, and most of these are retrovirus safe vectors.

18.2.1 Life Cycle of Retroviruses

A retrovirus contains a core of RNA as the genetic material contained within a protein coat (capsid), enclosed by an outer envelope. The viral RNA genome contains long terminal repeats (LTR) at the 5' and 3' ends carrying the promoter and termination site, respectively. In between are three coding regions - *gag* for viral core proteins, *pol* for the enzyme reverse transcriptase, and *env* for the outer envelope, and a non-coding region called psi (ψ) region (the packaging signal for directing the assembly of RNA in forming virus particles) (see Fig. 9.21).

During infection, the RNA genome of the retrovirus is injected into the cell, and converted to DNA by the enzyme reverse transcriptase. The viral DNA is then integrated into the host chromosomal DNA, as provirus. The integrated viral RNA is transcribed together with cellular transcription. The transcribed viral RNA also serves as mRNA for the synthesis of viral proteins. The viral RNA and proteins are assembled in a process called "packaging" to generate new viable retroviruses (see Section 9.4.4).

18.2.2 Construction of a Safe Retrovirus Vector

Retroviruses are infectious, and must be modified to be suitable for introduction of therapeutic genes. First, recombinant provirus DNA is constructed by deleting the viral genes in the provirus, replacing them with the therapeutic gene. The ψ region required for the assembly of RNA in the packaging, and the LTR regions for transcription initiation and termination, are retained in the vector.

The resulting recombinant provirus DNA is introduced into packaging cells. The recombinant provirus DNA directs the synthesis of RNA containing the therapeutic gene sequence and the ψ region. However, it lacks the viral proteins for assembly. The missing genes for viral proteins are provided by a ψ- helper provirus in the same packaging cell. The therapeutic RNA (from the recombinant provirus), and the viral proteins (from the helper provirus) are packaged into new viruses. These newly generated viruses are safe vectors that contain therapeutic RNA but no genes for viral proteins; they cannot regenerate new viruses (see Section 9.4.4).

18.3 Gene Treatment of Severe Combined Immune Deficiency

Gene therapy involves *ex vivo* manipulation of target cells. The general scheme is to isolate cells from the patient, and grow them in suitable culture media and conditions. The desired therapeutic gene construct is introduced into the cells via the use of retrovirus safe vectors. The infected cells are screened for the production of the therapeutic protein, propagated to sufficient quantity and introduced back into the patient.

The first human gene therapy trial involves the treatment of Severe Combined Immune Deficiency (SCID) caused by adenosine deaminase (ADA) deficiency. ADA is an enzyme essential for the breakdown of deoxyadenosine. Deficiency of this enzyme causes a build-up of the purine, which is preferentially converted to the toxic deoxyadenosine triphosphate in lymphocyte T cells, leading to damage of the immune system.

In September 1990, a clinical trial was conducted using retroviral-mediated vector to transfer the ADA gene into the T cells of two patients (Blaese et al. 1995. *Science* 270, 476-480).

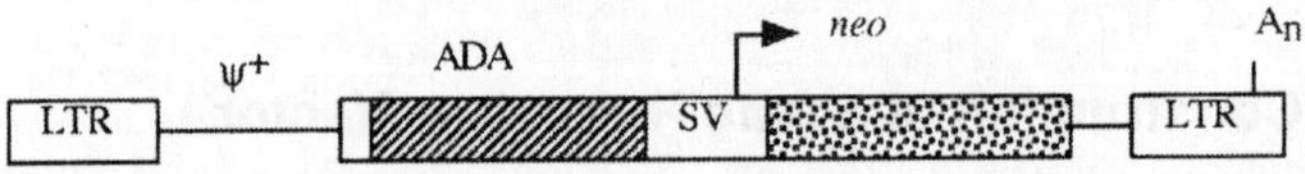

LTR = retroviral long terminal repeat containing the retroviral promoter and enhancer

ADA = Human adenosine deaminase cDNA under the Moloney murine leukemia virus (MoMLV) promoter.

SV = Simian virus 40 early region promoter

neo = neomycin phosphotransferase as a dominant selectable marker

ψ+ = retroviral packaging signal

An = Polyadenylation site

Fig. 18.2. The ADA retrovirus safe vector.

In the clinical trial, T cells were collected from the patient's blood, induced to proliferate in culture, in the presence of the hormone, interleukin 2. These cells were used for transfection with ADA retrovirus safe vector, and re-infused into the patient.

The retrovirus vector used to insert the ADA gene was derived from the Moloney murine leukemia virus (MoMLV) -based vector LNL6. The vector, known as LASN, contained the human ADA cDNA gene (1.5 kb) under the transcriptional control of the MoMLV promoter-enhancer in the retroviral LTR, and a *neo* gene controlled by an internal (simian virus 40 (SV40) promoter. The LASN was packaged with PA317 amphotropic retrovirus packaging cells (Fig. 18.2).

The LASN safe vector was used to infect the proliferating T cells in culture. The efficiency of gene transfer in cells ranged from 0.1 to 10% dependent on individual patients. The expression of ADA in the cells was monitored, and the transduction process was repeated with additions of the vector. The cultured cells carrying the ADA gene were then washed with saline containing 0.5% human albumin, and then infused into the patient. The gene treatment ended after 2 years, and the integrated vector and ADA gene expression in the patient's T cells persisted.

18.4 Therapeutic Vaccines

The discovery that "naked" DNA could be used as vaccines came about in the early nineties. It was found that plasmid DNA alone (without lipid carrier), when injected into the muscle of animals, was expressed *in situ*. Studies showed that the gene encoding the influenza virus antigens stimulate both specific humoral (antibodies and B cells) and cellular responses (cytotoxic T cells), accompanied by protection against a live influenza virus infection.

DNA-based immunization has since been shown to be effective in inducing protective immunity in various animal models, and may provide a potential alternative to traditional methods of vaccine development. Vaccines are currently developed by the use of live "attenuated" or "killed" bacterial and viral preparations. The former group includes vaccines for measles, mumps, and rubella, which stimulate both humoral and cell-mediated immune responses. The latter group includes vaccines for influenza, tetanus, hepatitis, which are relatively less potent, and primarily stimulate humoral immune response. DNA vaccination offers a molecularly defined, non-infectious route. For a number of infectious diseases, there are simply no effective vaccines. The ability to manipulate DNA vaccine antigens by recombinant means makes them particularly attractive to tackling infectious diseases for which there are no vaccination or drug treatment. Another target area of particular interest is the development of vaccines for cancer immunotherapy.

18.4.1 Construction of DNA Vaccines

DNA vaccines are mostly circular DNA plasmids, although other formats of nucleic acid vaccines can also be used. The construction of expression plasmids for producing antigens in mammalian cells requires the consideration of several key elements.

(1) A very strong promoter to insure maximum expression of the antigen. The human cytomegalovirus immediate early gene promoter (CMV) is used in most studies (See Section 9.4.2).

(2) The gene sequence of interest to be expressed. A Kozak sequence is needed upstream of the start codon. The stop codon is followed by a polyadenylation signal (See Section 5.5).

(3) A high copy number origin of replication to obtain high yield of the DNA preparation.

(4) Antibiotic resistance marker, such as kanamycin or neomycin for propagation and maintenance of the plasmid.

18.4.2 Methods of Delivery

In most DNA vaccination studies, the DNA is inoculated into skin or muscle, and the antigen expression occurs in keratinocytes and skeletal muscle cells, respectively. The antigen can be presented to the immune system in a manner similar to that after bacteria or virus infection. The methods of introducing the DNA into the subject is mostly done by intradermal or intramuscular injection. DNA vaccines can also be delivered using biolistic bombardment of DNA-coated gold beads by a gene gun similar to that use in the transformation of plant tissues (see Section 10.4).

Review

1. Describe the advantages and disadvantages of using physical/chemical methods and biological methods.
2. What is the role of a helper provirus in the construction of safe retrovirus vectors? What is the role of packaging cells?

3. What is LASN vector? List all the important elements and describe their functions.
4. In the clinical trial, why were proliferating T cells used for transfection with the ADA retrovirus safe vector, LASN?
5. What are the major elements in the construction of a plasmid used for DNA vaccines? Describe their functions relevant to the efficiency of expression and vaccination.

CHAPTER 19

GENE TARGETING

The delivery systems used in gene therapy are non-specific, infecting more than one cell type. In *ex vivo* or *in situ* manipulation this is not a serious problem. However, if *in vivo* therapy is to be developed, then cell specificity becomes desirable. In such a case, the gene carriers can be injected into the bloodstream much like administering many drugs.

19.1 Recombination

Gene targeting is a technology based on homologous recombination, a biological process occurred widely in prokaryotic cells and less frequently in eukaryotes. In homologous recombination, two double-stranded DNA molecules with a region of homologous sequence, line up adjacent to one another, and through a series of complex steps, exchange the two identical DNA segments. This type of homologous recombination, involving a swap of two homologous sequences, is known as reciprocal exchange or conserved exchange. In some cases, exchange of nucleotides in the homologous sequence may also be unidirectional. This type of exchange is non-reciprocal or non-conservative. It is also referred to as a gene conversion, because a portion of the recipient sequence is converted to the incoming sequence.

Homologous recombination provides a unique way to introduce foreign DNA into a specific location or to engineer genes *in situ* at their natural loci in the genome. Most gene targeting involves engineering alterations to a chosen gene for the purpose of studying gene structure/function. Targeted alterations of a chosen gene are called "gene knockout". The approach, however, is most appealing as a potential application in gene therapy. In the gene therapy protocols commonly in use, the introduction of a therapeutic gene integrates into the genome randomly, and thus requires transcriptional and translational regulatory elements in the gene construct. This is a complementation process in which a defective gene is augmented by introducing a functional gene. In contrast, gene targeting enables a direct replacement of a defective gene. The sequence carrying the mutation is replaced by the therapeutic gene sequence. The regulatory region of the gene may not need to be considered in the operation. Gene targeting, of course, has wide use and implication in other areas as well, such as in the production of transgenic plants and animals.

19.2 Replacement Targeting Vectors

There are several methods of constructing a vector for various selection purposes. In one of the original procedures, the engineered gene is constructed, so that the gene is interrupted by a selectable marker (e.g. *neo* gene), and flanked by short sequences homologous to the sequence in the genomic loci targeted for replacement. A second selectable marker (e.g. the thymidine kinase (*tk*) gene) is placed downstream of the gene and the homologous region. The two markers are known as positive and negative selectable markers, respectively. The entire construct (the gene plus homologous sequences at each end, plus selectable markers) is a replacement targeting vector (Fig. 19.1).

The vector is then introduced into suitable host cells by various methods, for example, microinjection, calcium phosphate precipitation, etc. Since the vector carries sequence homologous to the targeted gene in the genome, homologous recombination occurs replacing the genomic gene with the vector sequence. In homologous recombination, the vector aligns with the targeted gene in the chromosome. The segment of the vector carrying the engineered gene and the *neo* gene will replace the targeted gene, while the *tk* gene lying outside of the homologous sequence, will not be included in the replacement. At the same time, the majority of the recombination occurs in a non-homologous way, resulting in random insertion. In this case, the entire vector DNA (the replacement gene + *neo* gene + *tk* gene) will be incorporated into the cell chromosome at random.

The final step is to select cells containing the targeted replacement. This is achieved by a double selection by growing all the cells in a medium containing G418 and ganciclovir. Non-transformants will not survive because they do not carry the *neo* gene, and therefore sensitive to G418 (a neomycin analog). Cells resulting from non-homologous recombination carry the herpes virus *tk* gene and will be sensitive to the nucleoside analog, ganciclovir. The only cells that can grow in the medium are the ones generated by homologous recombination.

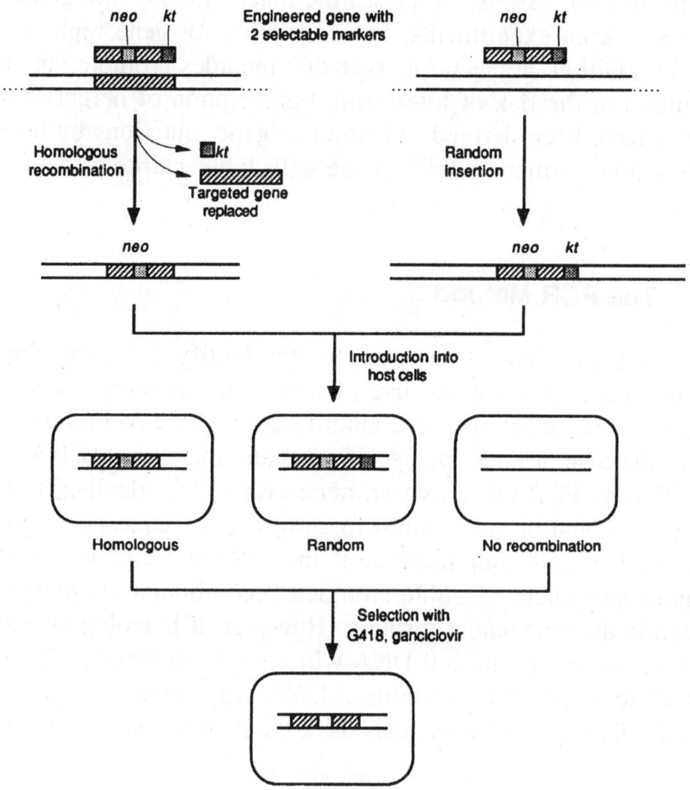

Fig. 19.1. Strategy in the use of a replacement targeting vector.

The procedure has been incorporated with the use of embryonic stem (ES) cells for potential gene replacement in live animals. The targeting vector is introduced into mouse embryonic stem cells in culture via homologous recombination as described. Stem cells are undifferentiated cells in the early stage of an embryo that gives rise to various cell types during development (see Section 22.1). The ES cells from the selection step are introduced into the embryo at the blastocyst stage. Since ES cells are capable of developing into many cell types, the resulting mouse will carry the mutation in various tissue cells, including germ cells. However, germ line transmission from transgenesis (generated by injection of ES-like cells into blastocysts) has not been demonstrated so far for any species except mice.

19.3 Gene Targeting Without Selectable Markers

The insertion of a selectable marker in a gene for targeting is not desirable for two reasons. It causes the inactivation of the gene, which is fine for knockout experiments, but unsuitable for gene replacement purposes. In addition, a genetic marker that includes promoter/enhancer elements may run the risk of interfering transcription of neighboring genes. Strategies have been derived to introduce gene mutations by homologous recombination, without retaining the selectable markers in the targeted loci.

19.3.1. The PCR Method

Strategies have been derived to identify cells carrying the replacement gene, without the use of selectable markers. The detection method is based on the selective amplification of the recombined DNA by PCR. In the case of targeting specific mutation to a gene, DNA from cells is amplified by PCR using two primers: primer 1 is identical to the mutation sequence, and primer 2 binds to an upstream sequence. Both primers will be used in PCR amplification if the cell DNA contains modified recombinant sequence. Double-stranded recombinant fragments will be generated in an exponential fashion. However, if homologous recombination has not occurred, the cell DNA will contain no binding site for primer 1, and PCR amplification yields ssDNA fragments non-exponentially. Modified cells are selected by analysis of the PCR products (Fig. 19.2).

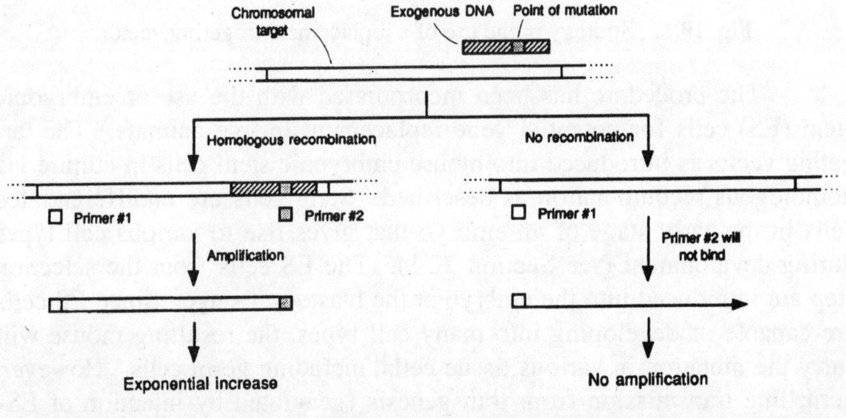

Fig. 19.2. Gene targeting without using selectable markers by PCR method.

19.3.2 The Double-Hit Method

In the double-hit gene replacement approach (also known as "tag and exchange"), two replacement type homologous recombination events are used. The first replacement vector is used to tag the gene by replacing part of the gene using positive (*neo* gene) and negative (*kt* gene) selectable markers. The resulting clones are subjected to positive selection (i.e. neomycin-resistance) to enrich for the replacement. In the second step, a replacement vector containing the gene with the mutation of interest is used to replace the selectable markers (*neo* and *kt*), and the clones that harbor the mutation can then be enriched by negative selection (Fig. 19.3).

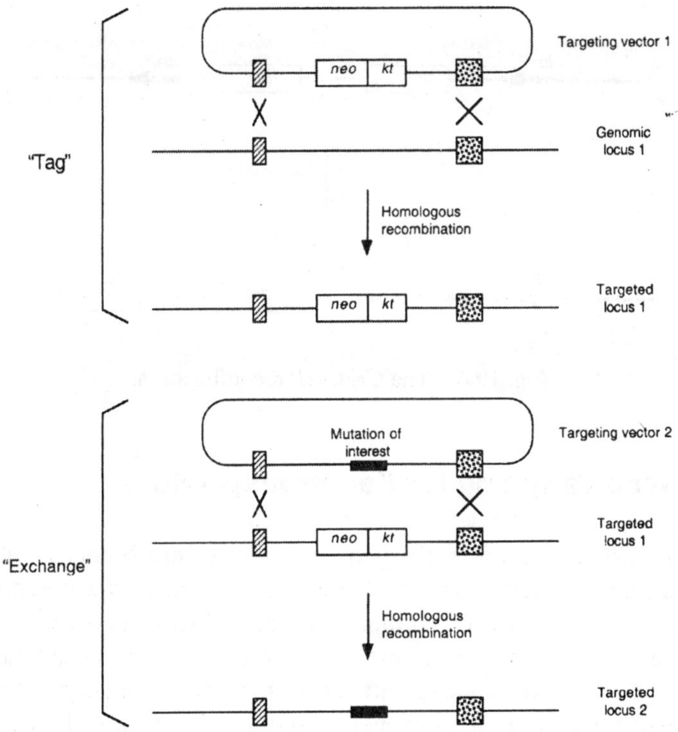

Fig. 19.3. Double-hit replacement.

19.3.3 The *Cre/loxP* Recombination

Another versatile strategy to introduce mutations is based on the *Cre/loxP* recombination system. The enzyme, Cre recombinase, recombines DNA at a specific DNA site containing a 34 bp sequence. This *loxP* site has two inverted 13 bp repeats separated by an 8 bp spacer. The enzyme catalyzes recombination resulting in the inversion of the intervening sequence when two *loxP* sites are arranged opposite orientation. The enzyme also catalyzes excision and recirculation of the intervening sequence when the two *loxP* sites are in the same orientation. In a general scheme, a replacement vector consisting of both positive and negative selectable markers flanked by two *loxP* sites and the desired mutation is inserted into the genomic locus of interest. In the second step, Cre recombinase is introduced to mediate excision of the markers, leaving one *loxP* site in the genome. The resulting clones that contain the introduced mutation can be enriched by negative selection (Fig. 19.4).

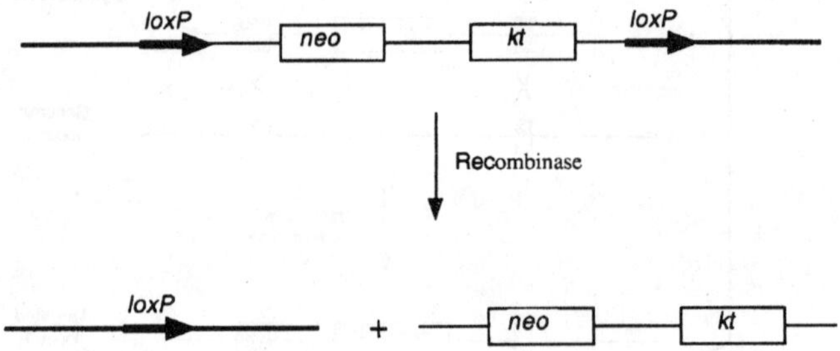

Fig. 19.4. The *Cre/loxP* recombination.

19.4. Gene Targeting for Xenotransplants

Transplanting animal organs and tissue into humans (xenotransplantation) has created much excitement and promise as a potential solution to deal with the severe shortage of human organs. Pigs are considered one of the better donors of organs, because they can be raised easily and their organs are similar in size and nature to those of humans. The major hurdle of using xenografts, however, is the development of hyperacute rejection and acute vascular rejection, resulting in the destruction of the grafts. The rejection is triggered by the binding of anti-donor antibodies in the recipient patient to the galactose-α1,3-galactose (α1,3Gal), a common carbohydrate moiety on the cell surface glycoproteins of almost all mammals, except humans, apes, and Old World monkeys. Since the key step in the synthesis of the α1,3-Gal epitope requires the enzyme α1,3-galactosyltransferase (α1,3GT), one of the approaches to eliminate the rejections is by knocking out the α1,3GT gene in the pig (Dai et al. 2002. *Nature Biotechnology* 20, 251-255).

A 6.4 kb α1,3GT genomic segment which expands most of exons 8 and 9 was generated by PCR from genomic DNA purified from porcine fetal fibroblast cells. The coding region of the pig α1,3GT gene is located in exon 9, and the gene is known to be expressed well in fetal fibroblasts. To create a targeting vector for the knockout of the α1,3GT gene, a 1.8 kb IRES-*neo*-poly A sequence was inserted into the 5' end of exon 9. The internal ribosome entry site (IRES) functions as the translation initiation site for the *neo* gene (which expresses the neomycin phosphotransferase protein as a G418 resistance marker). The *neo* gene has dual purposes of (1) disrupting the α1,3GT gene sequence and function, and also (2) providing a convenient screening strategy for positive clones based on G418 resistance (Fig. 19.5).

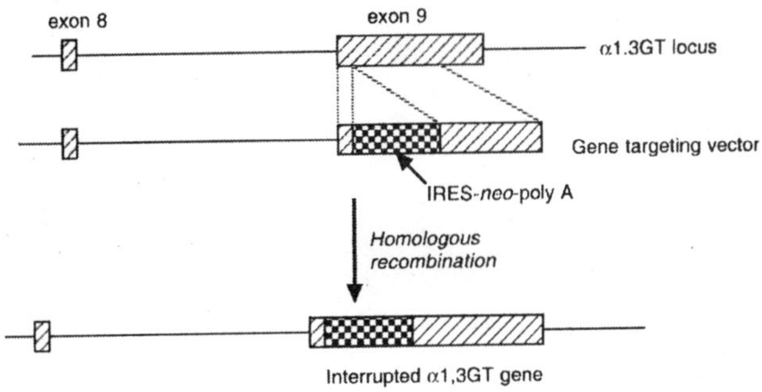

Fig. 19.5. Lockout of the α1,3GT gene.

The vector thus constructed was used to transfect cell lines derived from porcine fetal fibroblasts. Homologous recombination resulting in a knock-out α1,3GT gene was screened by recovering colonies that are resistant to G418. The insertion (knock-out) was further confirmed by PCR. In one of the transfected cell lines, 599 colonies were G418 resistant, 69 were confirmed by 3' PCR, and 18 were confirmed by long-range PCR. The 18 colonies were then subjected to southern blot to yield 14 positive colonies. Seven of the 18 Southern blot-confirmed α1,3GT knockout single colonies were used for nuclear transfer experiments to produce 5 female piglets of normal size and weight, all containing one disrupted pig α1,3GT allele. Starting with fibroblast cell cultures from such heterozygous animals, cells were selected in which the second allele of the gene was also mutated.

Review

1. What is homologous recombination? Reciprocal exchange? Nonreciprocal exchange?
2. What is "gene knockout"? What are the primary purposes of conducting such experiment?
3. How does a replacement targeting vector work?
4. What are the advantages and disadvantages of using selectable markers in gene targeting?
5. Describe one approach of gene targeting that does not require the use of selectable markers.
6. Describe one approach of gene targeting that does not retain the selectable marker.
7. In the knockout experiment, the *neo* gene was used to disrupt the α1,3GT gene. Why was the *neo* gene used for the experiment? Could point mutations be introduced to achieve the same purpose? Explain your answers.

CHAPTER 20

DNA TYPING

DNA typing (fingerprinting, profiling) has become one of the most powerful tools for paternity/maternity testing, criminal identification and forensic investigation. It is also an important tool in evolutionary studies of relatedness in animals, insects, and microorganisms.

20.1 Variable Number Tandem Repeats

The size of a human genome is ~3x10^6 bp, 2% of which are coding regions (exons) for the ~31,000 genes. The majority of the genome DNA (~98% including the introns) have no coding functions. Polymorphic (variable) markers that differ among individuals can be found throughout the non-coding regions.

The most useful for fingerprinting purposes are polymorphic markers containing repeated DNA sequences. These markers are typically defined by (1) the length of the core repeat unit, and (2) the number of repeats (i.e. the overall length of the repeat region). Variable number tandem repeat (VNTR) loci, also known as "minisatellites", contain 10-1000 repeats of 10-100 bp units. The number of repeats at a particular VNTR locus varies among individuals. This class of polymorphic markers has been extensively used for paternity analysis since its discovery in the mid-eighties. A few years after the discovery, short tandem repeat (STR) loci, containing 10-100 repeats of 2 to 6 bp units have also been identified. STR loci are also known as "microsatellites". This class of polymorphic markers has become the first choice in forensic typing, and in some instances, a replacement of VNTR markers for paternity analysis and other applications.

20.2 Polymorphism Analysis Using VNTR Markers

When sizable blood samples are available such as in the case of paternity analysis, a marker system based on restriction fragment length polymorphism (RFLP) is often used. When DNA is digested with a restriction enzyme that cuts at sites flanking the VNTR locus (but not within the repeats), the length of the DNA fragments produced will vary with individuals depending on the number of repeats in the locus. The unique length patterns of the restriction fragments provide a DNA fingerprint of an individual (Fig. 20.1).

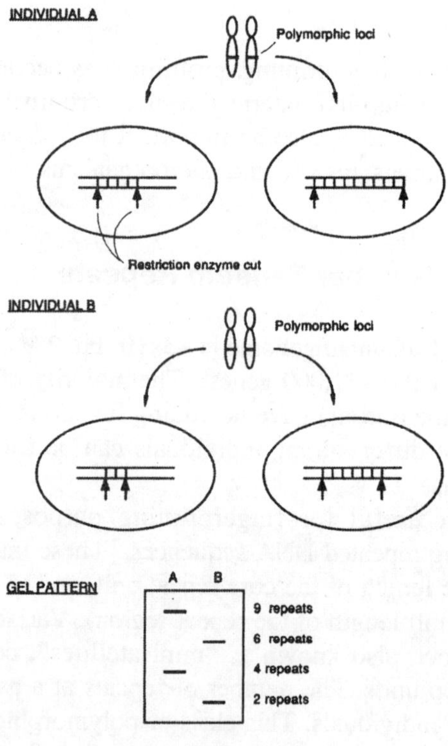

Fig. 20.1. The use of VNTR in generating fingerprints.

In practice, the DNA fragments obtained by restriction digestion are separated into bands according to their sizes by gel electrophoresis. The resolved bands are transferred by southern blot to a nitrocellulose membrane, which is subject to DNA hybridization using a radiolabeled probe having sequence complementary to the repeats of a VNTR locus (see below). The radiolabeled bands showing the DNA fingerprint are visualized by exposing the membrane to X-ray films by autoradiography (Fig. 20.2). Comparisons with patterns from known subjects can be achieved if parallel runs are made (see Sections 7.31 and 8.11 for non-radioactive detection).

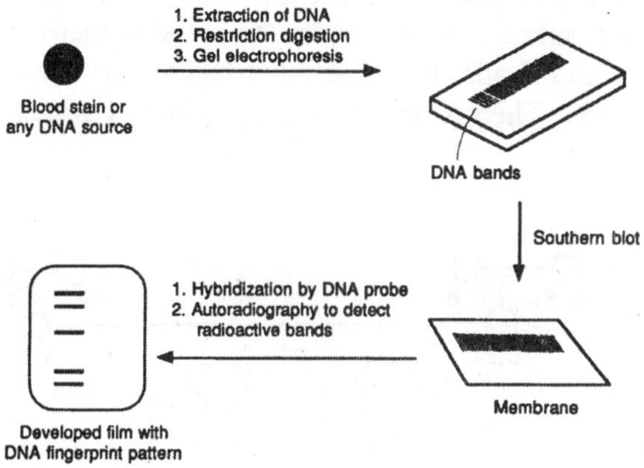

Fig. 20.2. Scheme of restriction fragment length polymorphism analysis.

20.3. Single-locus and Multi-locus Probes.

Two types of probes can be used: single-locus probes and the multi-locus probes. Single-locus probes detect a single locus in the genome, yielding patterns of two bands among the DNA fragments from restriction digestion and gel separation. Each band corresponds to each allele at the polymorphic locus in a homologous chromosomal pair. The commonly used restriction enzyme is *Hin*fI or *Hae*III. The frequently used single-locus probes include D1S7, D2S44, D4S139, D5S110, D7S467, D10S28, and D17S79. The designation of probes is based on chromosomal positions. The "D" stands for DNA, the number following refers to the chromosome number, the "S" refers to a single copy sequence, and the last number indicates the order the locus was discovered for a particular chromosome. In general, a combination of several single-locus probes is used. For a set of 5 probes, the probability of a random matching of unrelated samples is in the order of one in 10^{13} individuals. Multi-locus probes, simultaneously detect several loci that have some sequence similarities to permit hybridization to the same DNA probe. The widely used multi-locus probes, 33.6 and 33.15, detect 17 loci with DNA fragments consisting of 3-40 tandem repeats (2.5-20 kb range).

20.4 Paternity Case Analysis

The fingerprints of a three-generation family using a single-locus probe are shown in Fig. 20.3. The YNH24 (D2S44) probe reveals 8 alleles that can be followed unambiguously through the family tree, reflecting typical Mendelian inheritance. Each individual received one allele from one of his or her parents. For example, the grandmother was heterozygous having alleles 1 and 5, and the grandfather had alleles 2 and 7 (upper right corner in Fig. 20.3). Their daughter inherited alleles 1 and 2.

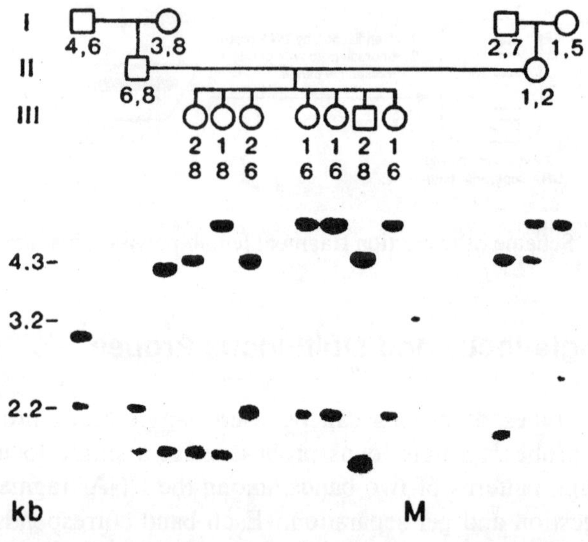

Fig. 20.3. Autoradiograms of Southern transfers from three-generation families. Single-locus probe pYNH24 was used. The genotypes of individuals in each three-generation family are shown directly below their symbols in the pedigree. (Reprinted with permission from Nakamura, Y., Leppert, M., O'Connel, P., Wolff, R., Holm, T., Culver, M., Martin, C., Fujimoto, E., Hoff, M., Kumlin, E., and White, R. 1987. Variable number tandem repeat (VNTR) markers for human gene mapping. *Science* 235, 1619. Copyright 1987, American Association for the Advancement of Science.)

If a combination of single-locus probes, or a multi-locus probe is used, the pattern of fingerprints becomes more complex, although the individual specificity is greatly enhanced. As shown Fig. 20.4 for a paternity case analysis, using the multi-locus probe 33.15, there were 22 scorable bands. The child had 6 of the maternal bands, 11 of the paternal bands, and 5 bands from either or both parents. There were 5 bands shared between both parents.

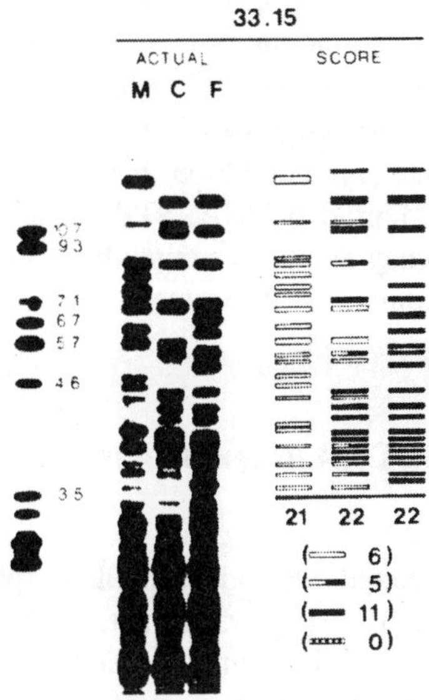

Fig. 20.4. Example of paternity case analysis. DNA figerprints were produced from Hinf1-digested genomic DNA prepared from the mother (M), child (C), and alleged father (F) using multilocus probe 33.15. The molecular weight marker was a restriction digest of single-locus minisatellite clones that were detected by the 33.15 probe. Sizes are in kb. The scoring analysis of the DNA fingerprints produced by each probe is shown schematically: maternal band; paternal band; child band from either or both parents; bands shared between parents; ●unassignable offspring band. Band scoring is limited to bands >3.5 kb. (Reproduced with permission from Jeffreys, A. J., Turner, M., and Debenham, P. 1991. The efficiency of multilocus DNA fingerprint probes for individualization and establishment of family relationships, determined from extensive casework. *Am. J. Hum. Genet.* 48, 821. Copyright 1991 by The American Society of Human Genetics, The University of Chicago Press.)

20.5 Short Tandem Repeat Markers

For forensic applications, short tandem repeat polymorphic markers are predominantly used because of several desirable characteristics. One major advantage is related to the fact that the total length of STR markers is significantly shorter than VNTRs, usually between 100-450 bp.

STR markers can be easily amplified and their shorter length also permits multiplexing (also known as multiplex PCR). Multiplexing is accomplished by using more than one primer set for the PCR reaction mixture, resulting in simultaneous amplification of two or more regions of DNA. The ability to perform PCR amplification and multipexing with STR markers means that a minute amount of sample DNA (0.1 to 1 ng), even in degraded form, can now be successfully typed. In contrast, RFLP methods require at least 0.1-0.5 μg non-degraded DNA. Considering that biological specimens in crime scenes (such as blood, hair, semen, etc.) contain very small amount and often degraded DNA, STR markers are the choice for forensic typing, because they are more compatible with the use of PCR. A plucked hair with root contains ~30 ng DNA, while a hair shaft contains only 0.1 ng of DNA.

20.5.1. The Combined DNA Index System.

The National Institute of Standards and Technology (NIST) has compiled a STR DNA Internet Database (www.csti.nist.gov/biotech/strbase). This STRBase provides details on all commonly used STR markers for the forensic DNA typing community. A core set of 13 STR markers is used to generate a nationwide DNA database in the United States, called the FBI Combined DNA Index System (CODIS) (Fig. 20.5). A parallel process of creating national DNA databases has been implemented earlier in several European countries.

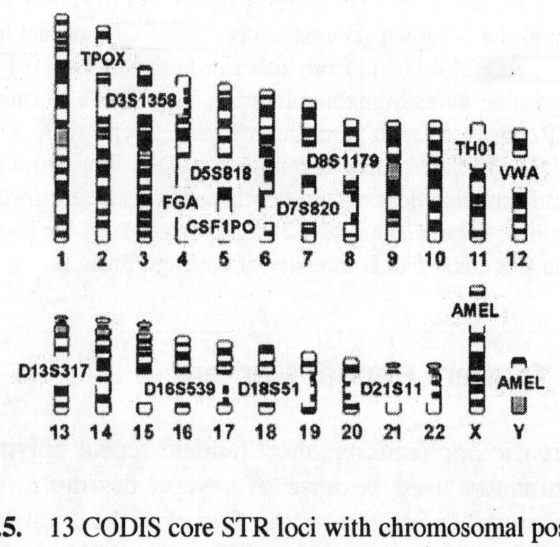

Fig. 20.5. 13 CODIS core STR loci with chromosomal positions.

The CODIS polymorphic loci are CSF1PO, FGA, TH01, TPOX, VWA, D3S1358, D5S818, D7S820, D8S1179, D13S317, D16S539, D18S51, D21S11 and Amelogenin. The last marker, Amelogenin, belongs to a group of sex-typing markers - It displays a 212 base X-specific band and a 218 base Y-specific band after amplification, and primarily used for gender identification. All 13 loci are highly polymorphic, and located outside of the coding regions. The probability of a match between two unrelated people is one in a trillion when all 13 CODIS loci are tested. Multiplex STR kits containing up to 10 STR loci are commercially available for used by forensic laboratories.

20.6 Mitochondrial DNA Sequence Analysis

In situations where nuclear (chromosomal) DNA typing is not an option (for example, insufficient quantities or too degraded), or an attempted typing using nuclear DNA markers is unsuccessful, mitochrondrial DNA (mtDNA) typing can be used.

Mitochondria have an extranuclear DNA genome, the sequence of which was first reported for humans in 1981. The human mtDNA is circular with 16,569 bp (as opposed to the linear ~3 billion bp in the nuclear DNA), and it exists in hundreds to thousands of copies in a single cell. The likelihood of recovering mtDNA from very minute and degraded biological samples is greater than for nuclear DNA. MtDNA has been extracted from teeth, hair shafts, bone fragments, all of which fail to yield forensic results with nuclear DNA markers. Most importantly mtDNA comes solely from the mother through the mitochondria in her egg, and therefore, represents only the maternal ancestry of an individual. Consequently, mtDNA information can reveal ancient population history and human evolution in anthropological investigations.

The forensic value of mtDNA lies in the displacement loop (D-loop) of about 1,100 bp in length located in the non-coding region. The two hypervariable regions (HV1 and HV2) of the D-loop is amplified by PCR, providing sequence information for positions 16,024-16,365 and 73-340, respectively (Fig. 20.6). The sequence can then be compared to the available forensic database of human mitochondrial DNA sequences. The non-coding region (also known as the control region) has been estimated to vary about 1-3% between unrelated individuals, with the variations distributed throughout the HV1 and HV2 regions. .

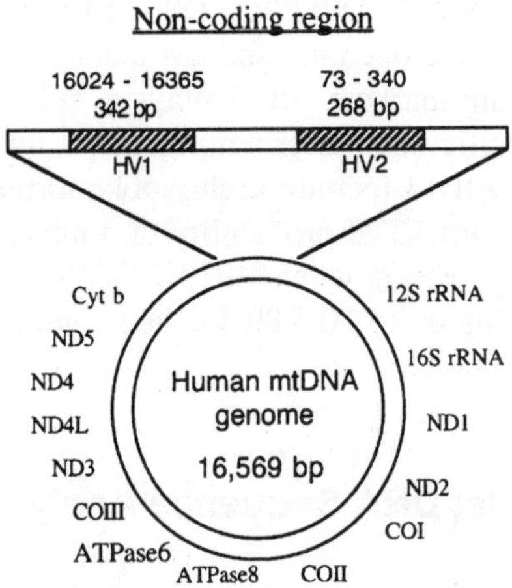

Fig. 20.6. The human mitochondrial DNA genome. The non-coding region includes the hypervariable regions, HV1 and HV2, used in DNA typing. Labeled genes: ATP synthase (ATPase), cytochrome c oxidase (CO), cytochrome b (cyt b), NADH dehydrogenase (ND).

The use of mtDNA sequence information to identify the 70-year old remains of the Russia Tsar Nicholas II and his family illustrates the power of DNA typing. In 1991, nine sets of skeletal remains were excavated from a shallow grave in Ekaterinburg, Russia, which were tentatively identified to include those of Nicholas II, Tsarina Alexandra, their three daughters, three of their servants, and the family doctor, Eugeny Botkin. In 1992, scientists at the UK and the United States were requested to collaborate in the verification of the remains using DNA based techniques (Gill et al. 1994. *Nature Genetics* 6, 130-135; Ivanov et al. 1996. *Nature Genetics* 12, 417-420).

The sex of the remains were determined by amplifying the amelogenin loci, and the results confirm the conclusion drawn from physical examination of the bones that the remains include four males and five female bodies. The nine skeletons were further subjected to DNA typing using 5 STR markers: VWA/31, Thol, F13A1, FES/EPS, and ACTBP2. The allele band patterns indicate that 5 of the skeletons belonged to a family group, comprising two parents and three children.

The method of mtDNA typing was used to compare the sequence of the putative Tsar with two living maternal descents from the Tsar's maternal grandmother – The great great-grandson (Duke of Fife), and the great-great-great-granddaughter (Countess Xenia Cheremeteff-Sfiri) of Louise of Hesse-Cassel (the Tsar's grandmother). The putative Tsarina's and the three daughters' mtDNA sequences were compared with that of Prince Philip, Duke of Edinburgh, a grand-nephew of maternal descent from Tsarina Alexandra.

The mtDNA sequences of the putative Tsarina and the three daughters were the same with that of Prince Philip, confirming the identities of the mother and the siblings. The mtDNA sequence extracted from the putative Tsar's bones matched with those of the two maternal relatives of Nicholas II, with the exception at one position. At base position 16,169, the DNA sequence revealed the presence of both a cytosine (C) and a thymine (T) at a ratio of 3.4:1, while the two maternal relatives contained only T at this position.

The sequencing discrepancy was resolved by a following analysis of the exhumed remains of the Grand Duke of Russia, Georgij Romanov, who was the brother of Tsar Nicholas II. Georgi Romanov had a matching mtDNA hypervariable sequence with the same C/T heteroplasmy at position 16,169, with a ratio of 38% C and 62% T. (The presence of more than one mtDNA type in an individual is known as heteroplasmy, resulting in more than one base at a site in the mtDNA sequence.) Thus, the previous discrepancy observed between the putative Tsar and the two relatives was due to heteroplasmy passed from the Tsar's mother to the two sons – Georgij and Nicholas, but segregated to homoplasmy in genetic transmissions during succeeding generations. The authenticity of the remains of the Romanov family has finally been confirmed.

Review

1. What are polymorphic markers? Why are these markers suited for fingerprinting?
2. What are the differences between VNTR and STR?
3. Why are STR markers preferred for forensic typing?
4. Why are *Hin*fI and *Hae*III the common choice for restriction enzymes? (Hint: How critical is it to control the fragment size in runing restriction fragment length polymorphism analysis?)
5. What is the reason for choosing the 13 loci in CODIS?
6. What is multiplexing? Describe the role of PCR multiplexing in DNA typing.
7. Follow the band patterns in Fig. 20.3, and confirm that they reflect Mendelian inheritance.
8. What are the differences, advantages and disadvantages between single-locus and multi-locus probes?

9. What regions of the mtDNA genome are used for DNA typing? Explain why these regions are used.
10. Human genome DNA in the nuclear and in the mitochondria shows different characteristics.

	Nuclear DNA	Mitochrondria DNA
Size:		
3 billion bp or16569 bp		
Copies per cell:		
>100 or 2		
Inherited from:		
mother or both parents		
Recombination rate:		
high or low		

CHAPTER 21

TRANSPHARMERS - BIOREACTORS FOR PHARMACEUTICAL PRODUCTS

The application of transgenic technology to commercially important livestock is expected to generate major effects in agriculture and medicine. Three areas of development have been the focus of intensive investigation: (1) For improved desirable traits, such as increased growth rate, feed conversion, reduction of fat, improved quality of meat and milk. Growth hormone transgenes have been inserted into genomes of pig, sheep, and cow; (2) For improved resistance to diseases - A number of genes contributing to the immune system (such as heavy and light chains of an antibody that binds to a specific antigen) can be introduced to confer in *vivo* immunization to transgenic animals; (3) To raise transgenic animals for the production of pharmaceutical proteins - The concept of using farm animals as bioreactors has raised the prospect of a revolutionary role of livestock species. The list of proteins includes human lactoferrin, human collagen, α_1-antitrypsin, blood coagulation factor, anticlotting agents, and others.

The prospect of producing pharmacologically active proteins in the milk of transgenic livestock is appealing for several reasons. (1) Transgenic animals may ultimately be a low-cost method of producing recombinant proteins than mammalian cell culture. Lines of transgenic livestock, although are costly to establish, can be multiplied and expanded rapidly and easily. In contrast, the maintenance of large-scale mammalian cell culture requires continuous high expense. (2) Unlike microbial systems that are not capable of posttranslational processing, transgenic animals produce bioactive complex proteins with an efficient system of post-modification. (3) Recovery and purification of active proteins from milk is relatively simple. The volume of milk production is large, and the yield of protein may be potentially high, rendering the process economically feasible.

21.1 General Procedure For Production of Transgenic Animals

In a general scheme, the gene of a desired protein is constructed in a suitable vector carrying the regulatory sequence of a milk protein which to direct the expression in mammary tissues. Promoters that have been used often include those of the genes of β-lactoglobulin and β-casein (major proteins found in milk). The recombinant DNA is then introduced into the pronuclei of fertilized eggs at an early stage by microinjection. The injected DNA is usually integrated as multiple tandem copies at random locations. The transformed egg cell is then implanted into the uterus of a surrogate animal to give birth to transgenic offspring. The transgenic animal can be raised for milking the expressed protein for processing and purification. Stable transmission of the transgene to succeeding generations is a critical factor in establishing transgenic lines of the livestock. Although it is not as frequent, transgenes can also be introduced using nuclear transfer techniques (see Sections 22.2).

21.2. Transgenic Sheep for α_1-Antitrypsin

The raise of transgenic sheep for the production of α_1-antitrypsin has been described (Wright et al. 1991. *Bio/Technology* 9, 830-834). Human α_1-antitrypsin (Hα_1AT) is a glycoprotein with a molecular weight of 54 kD, consisting of 394 amino acids, with 12% carbohydrates. The protein is synthesized in the liver and secreted in the plasma with a serum concentration of ~2 mg per ml. Human α_1AT is a potent inhibitor of a wide range of serine proteases, a class of enzymes, if leave unchecked, can cause excessive tissue damage. Individuals deficient in the protein risk the development of emphysema.

In the study, a hybrid gene was constructed by fusing the Hα_1AT gene to the 5' untranslated sequence of the ovine β-lactoglobulin (βLG) gene. The Hα_1AT gene consisted of five exons (I, II, III, IV, and V) and four introns. In the gene construct, the first Hα_1AT intron (between exons I and II) sequence was deleted. This Hα_1AT minigene therefore consisted of exons I and II fused, and exons III, IV and V interrupted by introns, the

Hα_1AT initiation codon (ATG), stop codon (TAA), and polyA termination signal. The 5' untranslated βLG sequence included the βLG promoter, the TATA box, and the βLG exon I sequence (Fig. 20.1).

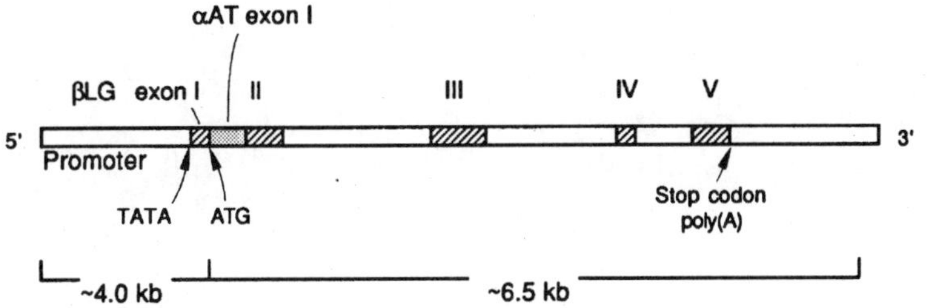

Fig. 21.1. The hybrid gene construct of human α_1-antitrypsin fused with the 5' untranslated sequence of the ovine β-lactoglobulin gene.

The hybrid gene construct was microinjected into sheep eggs collected from donor ewes following artificial ovulation and insemination. Southern blot analysis of the genomic DNA samples identified 5 transgenic animals from 113 lambs. The transgene was shown integrated in multiple (2-10) copies. Three of the transgenic sheep produced offspring, and these three lactating sheep were used for daily milk collection. The milk samples were analyzed by radial immunodiffusion assay for the presence of Hα_1AT. The milk samples were also used to purify the protein for sodium dodecyl sulfate – polyacrylamide gel electrophoresis (SDS-PAGE) analysis. All three transgenic sheep produced the human protein exceeding 1 g per liter. The protein appeared to be glycosylated and fully active.

Review

1. List the advantages and disadvantages of using livestock animals for the production of pharmaceutical proteins.
2. Why are promoters of the β-lactoglobulin and β-casein genes used for animal transgenes?
3. In the example described, the transgene was integrated in multiple copies in the genome.
4. Can a transgene be integrated by targeting a specific location in the chromosome? Explain your approach.

CHAPTER 22

ANIMAL CLONING

A revolutionary event in biology and medicine occurred in 1996 when scientists at the Roslin Institute in Scotland succeeded in cloning animals from cultured cells taken from a mature ewe. Dolly is the first mammalian clone created by transferring the nucleus from an adult cell to an unfertilized egg (with its own nucleus already been removed). Clones have since been produced from adult cells of mice, cattle, goats, pigs and other animals.

22.1 Cell Differentiation

Fertilization of an egg by a sperm results in the formation of a zygote which ultimately gives rise to all the cells of the adult body – more than a hundred trillion cells of diverse structures and functions – through progressive developmental changes. The zygote begins the process of cleavage in which it undergoes rapid division from a single cell to 2 daughter cells, 4, 8, 16, and so on. During cleavage, the embryo retains approximately the same overall external spherical form with little change of the overall volume. This means that the daughter cells (known as blastomeres) become smaller and smaller with each cell division. The cleavage process ends with the formation of a hollow structure called the blastula, with the blastomeres moved to the periphery leaving a fluid-filled cavity in the center. As the embryo enters into this stage, the cells become differentiated.

Differentiation is a process whereby originally similar cells follow different developmental pathways into specialized cells, for example, nerve cells, muscle cells, etc. – that eventually make up the various tissues and organs of the body. And the process is all controlled by the collective actions of genes in a particular group of cells. The cells produced from the first few divisions after fertilization are undifferentiated, meaning that they can develop into any of the cell types. Undifferentiated early embryonic cells have been used as the source of choice for cloning using nuclear transfer techniques, prior to the cloning of Dolly.

22.2 Nuclear Transfer

In performing nuclear transfer, the nucleus is first removed from an unfertilized egg (oocyte) taken from an animal soon after ovulation. This is accomplished by using a dedicated needle to pierce through the shell (zona pellucida) to draw out the nucleus under a high power microscope. The resulting cell, now devoid of genetic materials, is an enucleated oocyte. In the next step, the donor cell carrying its complete nucleus is fused with the enucleated oocyte. The fused cells then develop like a normal embryo, and finally implanted into the uterus of a surrogate mother to produce offspring. Instead of using a whole donor cell to fuse with the recipient cell, the donor cell nucleus can be removed and transferred by injecting the DNA directly into the recipient cell.

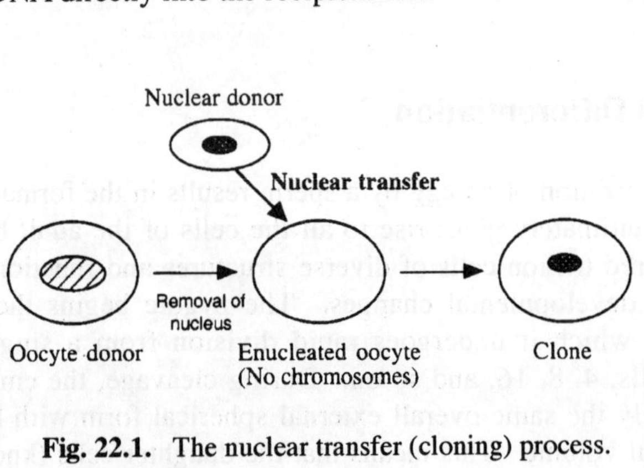

Fig. 22.1. The nuclear transfer (cloning) process.

The technique of nuclear transfer was first applied to cloning frogs in 1952, but the cells never developed beyond the tadpole stage. In the mid-eighties, several research groups succeeded in producing sheep and cattle by nuclear transfer using early embryonic cells. In some later studies, cells from embryos that had advanced to 64- and 128-cell stages were used to produce calves.

The major breakthrough that set the stage for creating Dolly came in 1995 when scientists at the Roslin Institute successfully produced lambs by nuclear transfer from cells taken from early embryos that had been *cultured* for several months in the laboratory. The experiment using cultured embryonic cells led to the cloning of Dolly using adult (differentiated) cells, which sets it apart from all previous cloning attempts of employing embryonic (undifferentiated) cells. The success of cloning adult cells proves that cell differentiation is reversible, and the hands of time in a developmental process can be manipulated to reprogram its course.

22.3 The Cloning of Dolly

(Wilmut, I., et al. 2002. *Nature* 419, 583-586.) The key to the success of cloning Dolly was the careful coordination of the cell cycle of the donor cell. The cells used as donors for nuclear transfer were at a stage of the cell cycle called "quiescence" – the state at which the cell is arrested and stops dividing

For most cells, the life history can be represented as a repeating cycle of metaphase (M phase, see Section 1.5 on mitosis) and interphase. The cell's DNA replicates during a special portion of the interphase, called the S phase. The interphase also contains two time gaps: G1 between the end of mitosis and the start of the S phase, and G2 separating the S phase and onset of mitosis. No DNA is made during G1 and G2, however, protein synthesis occurs throughout the entire interphase.

For simplicity, one can view a cell cycle to consist of two major phases – one phase for nuclear division to form two daughter cells (mitosis), and another phase for DNA and protein synthesis. Cells that do not divide are usually arrested in the G1 phase.

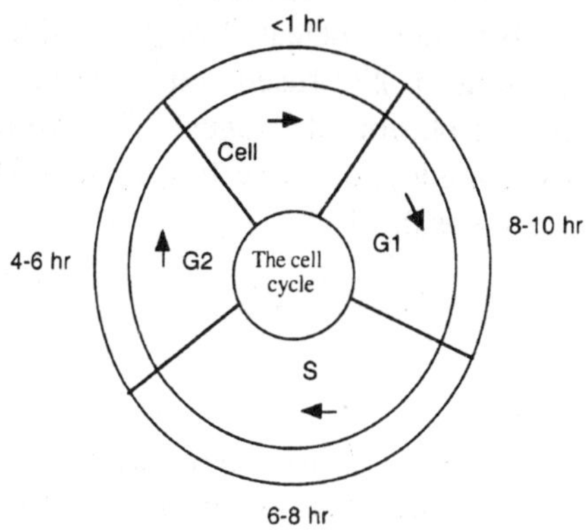

Fig. 22.2. The cell cycle.

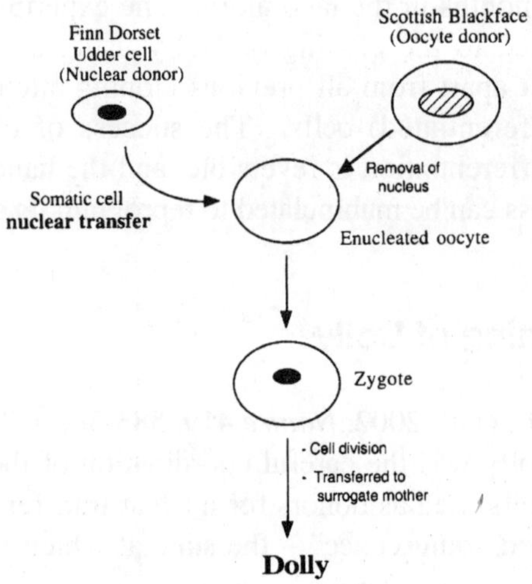

Fig. 22.3. The cloning of Dolly.

For the cloning of Dolly, the donor cells used were derived from the mammary gland of a 6-year old Finn Dorset ewe in the last trimester of pregnancy. The cells were made quiescent by growing in media with reduced concentrations of nutrients for 5 days. Unfertilized eggs were obtained from Scottish Blackface ewes between 28 to 33 hr after injection of gonadotropin-releasing hormone, and the nuclei were removed. The following step was to transfer the nucleus from a quiescent mammary donor cell into the enucleated oocyte. Electrical pulses were applied to induce fusion of the donor cells to the enucleated oocyte and to activate the cell to development. From a total of 277 fused cells obtained, 29 developed into embryos (at the morula or blastocysts stages). They were implanted into 13 ewes, resulting in 1 pregnancy, and one live lamb – Dolly. The success rate of the entire process was 0.4%.

Review

1. Describe the general scheme in the nuclear transfer?
2. Why have early embryonic cells been commonly used for animal cloning?
3. What is unique about the technique used in cloning Dolly that distinguishes it from previous animal cloning experiments?

CHAPTER 23

HUMAN GENOME SEQUENCING

The publication of the human genome sequence is the cumulative result of nearly five decades of international collaborations. The Human Genome Project's (HGP) sequencing strategy is a clone-by-clone or hierarchical strategy, first producing genetic and physical maps of the human genome (the first 5-year plan, 1993-1998), and then pinning the sequences to the genome map (the second 5-year plan, 1998-2003).

The first phase of the HGP has focused on mapping the human genome. Mapping is the construction of a series of chromosome descriptions that depict the position and spacing of unique, identifiable biochemical landmarks that occur on the chromosomes. These landmarks, known as "DNA markers" in a genome can be specified according to the relative positions based on the use of (1) genetic techniques including linkage analysis of polymorphic markers (genetic mapping) and (2) direct physical analysis of distinctive sequence features in the DNA molecule (physical mapping). The resulting genome map is a comprehensive integration of genetic and physical maps that ultimately provides the framework for carrying out the sequencing phase of the project.

23.1 Genetic Maps

A genetic (linkage) map is a description of the relative order of genetic markers in linkage groups in which the distance between markers is expressed as units of recombination (calculation based on meiotic recombination). It orders and estimates distances between markers that vary between parental homologues (polymorphism) by linkage analysis. The closer the markers are to each other, the more tightly linked, the less likely a recombination event will fall between and separate them. Recombination frequency thus provides an estimate of the distance between two markers. The primary unit of distance along the genetic map is the centi-Morgan (cM), which is equivalent to 1% recombination. A genetic distance of 1 cM is approximately equal to a physical distance of 1 million bp (1 Mb). The basic principles of genetic linkage have been described in connection with Chapter 17: Finding Disease-Causing Genes) .

23.1.1 DNA Markers

The early genetic maps were constructed based on Mendelian genetics, by observing the changes of heritable characteristics, and thus the changes in phenotypes displayed in the offspring as compared to the parents. This type of mapping technique relies on visualization of a particular phenotype and requires extensively planned breeding experiments. Phenotype observation has its complications, because a single physical feature is often controlled by more than one gene.

This method has been largely replaced by the use of DNA markers that can be studied by biochemical techniques. The first type of DNA markers contains mutations that cause changes in a restriction site sequence, which becomes not recognized by the corresponding restriction enzyme. These markers can be detected by restriction fragment length polymorphism (RFLP) analysis. A restriction cut by the enzyme produces a longer DNA fragment because the two adjacent restriction fragments remain linked together. Genetic markers are typed by hybridization or PCR. The region surrounding the marker sequence is amplified, the DNA is treated with restriction enzymes, and the fragments are separated according to their sizes by agarose gel electrophoresis. The positions of the bands on the gel correspond to the length of the amplified fragment, and therefore reveal the state of polymorphism (see Section 20.2).

In fact, any polymorphic lòci that are unique to the genome can be used for genetic mapping. For example, short tandem repeats, AC/TG repeats, and tri- and tetranucleotide repeats (see Sections 20.1 and 20.2) are popular DNA markers. Single nucleotide polymorphism (SNP), individual point mutations occurred abundantly in the genome sequence, can also be used in genetic mapping.

23.1.2 Pedigree Analysis

Linkage analysis with humans is quite different from that of other organisms. For example, in the study of fruit flies or mice, extensive breeding experiments can be planned and designed for gene mapping purposes. However, planned experimentation to select crossings is impossible with humans. Instead, the data in humans are limited to those that can be collected from successive generations of an existing family, hence the term "pedigree analysis". For this reason, family collections have been established, and are accessible to researchers for marker mapping. An example is the collection maintained by the Centre d'Etudes du Polymorphisme Humaine (CEPH) in Paris, which consists of cultured cell lines from eight families comprised of a total of 134 individuals providing 186 meiotic recombinations. The collection enables investigators world wide to employ a common set of families and pool data from markers developed in individual laboratories. In human genetic analysis, because the number of genotypes is small and the nature of the pedigree is imperfect, the markers are analyzed statistically by the use of a lod (logarithm of the odds) score. A lod ratio >1000:1 (an odds ratio of at least 1000:1 against alternative orders) is taken as significant, suggesting the markers are linked. Marker mapping is now performed entirely by computer-based analysis tools.

23.2 Physical Maps

Physical maps are used to isolate and characterize unique DNA sequences, including individual genes, and provide the substrate for DNA sequencing.

23.2.1 Sequence Tagged Sites

Physical mapping of the genome relies on markers generally called "sequence tagged sites (STS)". Any short stretches of sequence (typically less than 500 bp) can be used as STS provided: (1) It has a unique location in the chromosome; (2) Its sequence is known so that it can be detected by PCR assays. A common source of STS is the expressed sequence tag (EST). EST is obtained by performing a single raw sequence read from a random cDNA clone. Since cDNAs are obtained from reverse transcription of corresponding mRNAs, random EST sequencing is a rapid means of discovering sequences of important genes even the expressed sequences are often incomplete. The use of EST in STS mapping has the advantage that the mapped markers would correspond to coding regions in the genome.

The primary objective of mapping is to enable integration of physical and genetic mapping data across chromosomal regions. These maps will facilitate the construction of a comprehensive, integrated platform for sequencing and identification of disease genes.

23.2.2 Radiation Hybridization

To perform an STS mapping, one needs to generate a collection of DNA fragments spanning a human chromosome or the entire genome. This collection of DNA fragments is called "mapping reagent". One approach to obtain such collections is using radiation hybrid (RH) panels. A RH panel consists of many large fragments of human DNA produced by radiation breakage fused in hamster fibroblast cell lines. To create RH panels, human cells are exposed to X-ray radiation to randomly fragment the chromosome, and then fused with hamster cells to form hybrids that can be propagated as cell lines (Fig. 23.1).

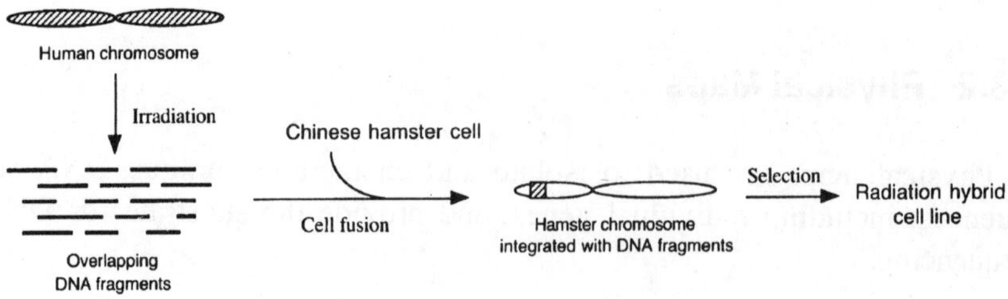

Fig. 23.1. Radiation hybrids.

To type an STS, a PCR assay is used to score all the cell lines in a panel for the target STS sequences. The frequency of detected two STS markers in the same fragment depends on how close they are together in the genome. The closer they are, the greater chance will be for both detected on the same fragment. The further apart they are on the genome DNA, chances are less likely for them to be found on the same fragment. The physical distance is based on the frequency at which breaks occur between two markers.

23.2.3 Clone Libraries

A clone library can also be used as the mapping reagent for STS analysis. Libraries commonly used for this purpose are yeast artificial chromosome (YAC) and bacterial artificial chromosome (BAC) libraries.

The markers are assembled by detection of physical overlaps among the set of clones in the library with overlapping individual fragments (clone contigs). This is usually done by fingerprinting methods, such as cross-hybridization or PCR of genome sequence repeats or STS markers. For example, if PCR is directed at individual STSs with each member of a clone library, then all clones that give PCR products must contain overlapping inserts (Fig. 23.2). YAC has been initially used because it can accommodate large fragments and thus cover large distances. (Chapter 8, Section 17.2.3). However, YAC as well as cosmid are multi-copy vectors, and suffer from low transformation efficiencies, difficulty in getting large amount of insert DNAs from transformed cells, and instability problems of rearrangement and recombination. For these reasons, BAC has become the preferred vector to use in construction of contiguous libraries.

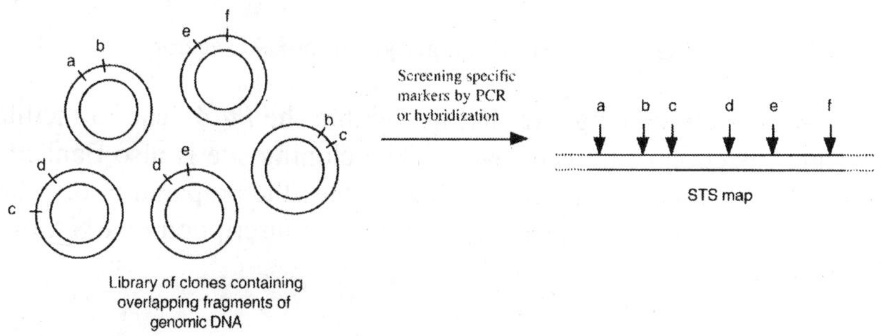

Fig. 22.2. The use of clone libraries as a mapping reagent.

23.2.4 The Bacteria Artificial Chromosome Vector

BAC is derived from the F factor of *E. coli*, which naturally occurred as a 100 kb molecule. It enables cloning of large (>300 kb) inserts of DNA in *E. coli*, which can be maintained at a very low copy number, approximately one copy per cell, a feature that favors stable replication and propagation (Shizua, H., et al. 1992. *Proc. Natl. Acad. Sci. USA* 89, 8794-8797). A typical BAC vector, such as pBeloBAC11, contains the following major features (Fig. 22.3)

(1) Sequences needed for autonomous replication, copy-number control, and partitioning of the plasmid, including *oriS*, *repE*, *parA*, *parB*, and *parC*, all derived from the F factor of *E. coli*.

(2) The chloramphenicol resistance gene as a selectable marker, and also two cloning sites (*Hin*dIII and *Bam*HI) and other restriction sites for potential excision of the inserts.

(3) Bacteriophage λ *cosN* site and the bacteriophage P1 *loxP* site. The *cosN* site can be conveniently cleaved by bacteriophage α terminase (a commercially available enzyme) to linearize the DNA. The *loxP* site is recognized by the Cre recombinase, and can be used to introduce additional DNA elements into the vector using Cre-mediated recombination process (see Section 19.3.3).

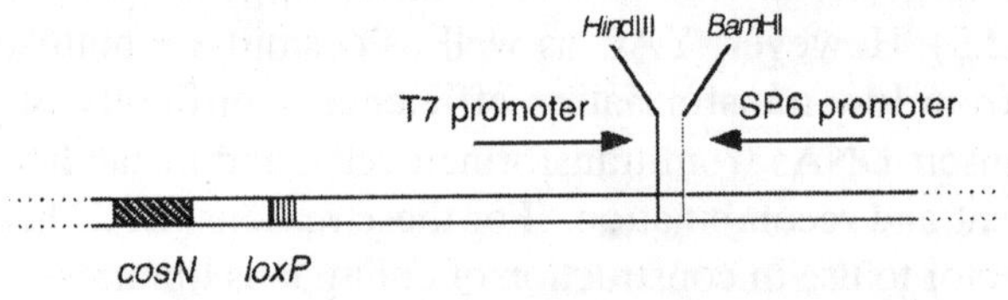

Fig. 23.3. The cloning segment of BAC vector.

(4) A multiple cloning site that lies within the *lacZ* gene to facilitate blue/white selection of transformants. The cloning site is also flanked by SP6 and T7 promoter sequences, which enables the preparation of probes from the ends of cloned sequences by *in vitro* transcription of RNA or by PCR methods.

23.3 Comprehensive Integrated Maps

High-density human genetic and physical maps started to emerge in the early nineties. Genetic (linkage) maps are based on polymorphic markers, such as short tandem repeat, AC/TG repeats, and tri- and tetranucleotide repeats. High-density physical maps are based on STSs by radiation hybridization and YAC/BAC cloning.

For example, a high-density genetic (linkage) map achieved in 1994 consists of 5,840 loci, 3,617 of which are PCR formatted short tandem repeat polymorphisms, and another 427 of which are genes, with an average marker density of 0.7 cM (Murray et al. 1994, *Science* 265, 2049-2054)). The 1996 version of the human linkage map consists of 5,264 short tandem $(AC/TG)_n$ repeat polymorphisms, with average interval size of 1.6 cM (Dib et al. 1996, *Science* 380, 152-154).

A physical map achieved in 1995 contains 15,086 STS markers, and later supplemented with 20,104 STS mostly ESTs, with a density of 1 marker per 199 kb (Hudson et al. 1995. *Science* 270, 1945-1954). A physical map released in 1998 assembles 41,664 STSs by RH mapping, with 30,181 of the STSs based on 3' untranslated regions of cDNAs representing unique genes (Deloukas et al. 1998. *Science* 282, 744-746).

The markers developed by genetic and physical mapping were compared and integrated to build the framework of a consensus map for the DNA sequencing phase of Human Genome Project. For the coordination of genome sequencing, each participating laboratories or centers in the international consortium have responsibility to completing one or more sections (minimum size of ~1 Mb) of the genome. Boundaries of the sections are defined by selection of unique markers from the framework.

23.4 Strategies For Genome Sequencing

The sequencing phase proceeds by an approach general known as "hierarchical shotgun sequencing strategy", also referred to as "map-based", or "clone-by-clone" approach. An alternative is the "whole-genome shotgun sequencing strategy".

23.4.1 Hierarchical Shotgun Sequencing

This approach involves two stages of cloning (Fig. 23.4). The overall scheme involves breakdown of the chromosomes into manageable large fragments, which are then physically ordered and sequenced individually by shotgun sequencing.

First, the genome is broken into manageable segments of 50 to 200 kb in size using partial digestion or sonic shearing. The fragments are inserted into the BAC vector, followed by transformation into *E. coli*, to create a library of clones covering the genome. The DNA fragments represented in the library are arranged and positioned at correct locations on the genome map constructed in the first phase of the HGP. In the second step, individual clones are selected and sequenced by random shotgun strategy. The DNA fragment in a clone is sheared into small fragments (2 to 3 kb) by sonication. These small DNA fragments are subcloned into plasmids or phagemids for sequencing.

The final step is to assemble a draft genome from individually sequenced BAC clones, by first ordering contigs for each BAC clone, and then aligning overlaps at the ends of BAC sequences that are adjacent to each other. All genome-scale sequencing has been performed with high-throughput automation, and sequence assembly achieved by the use of sequence editing software.

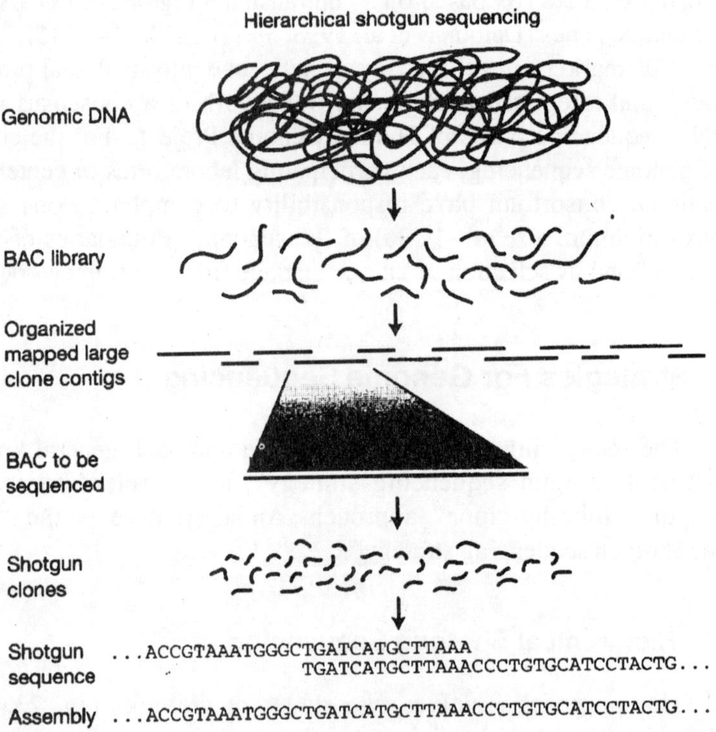

Fig. 23.4. The hierarchical shotgun sequencing strategy. (Reproduced with permission from International Human Genome Sequencing Consortium et al. *Nature* 409, 863. Copyright 2001 by Nature Publishing Group.)

23.4.2 Whole-Genome Shotgun Sequencing

An alternative strategy is known as "whole-genome shotgun sequencing strategy" or "direct shotgun sequencing strategy" (Fig. 23.5). This approach involves randomly breaking the genome into small DNA segments of various sizes (2 to 50 kb) and cloning these fragments to generate plasmid libraries. These clones are then sequenced from both ends of the insert. Computer algorithms are used to assemble contigs derived from thousands of overlapping small sequences. Contigs are connected (ordered and oriented) into scaffolds, and anchored onto chromosomal locations by with reference to the HGP mapping information. The BAC, STS, and EST sequence data derived from the clone-based strategy are utilized for sequence assembly and validation analysis.

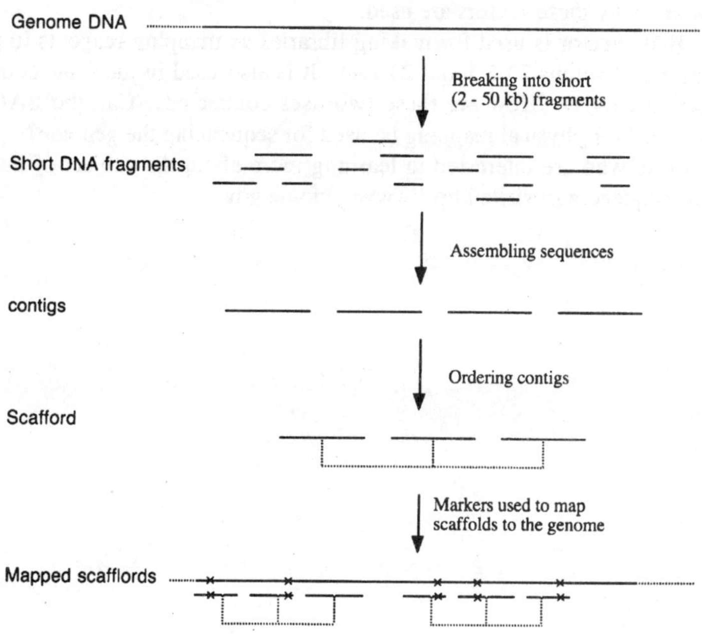

Fig. 23.5. The whole-genome shotgun sequencing strategy. (Adapted with permission from Venter, J. C., et al. The sequence of the human genome. *Science* 291, 1309. Copyright 2001 by American Association for the Advancement of Science.)

Review

1.

	Genetic map	Physical map
Markers used		
Unit distance of markers		
Methods of detecting markers		

2. Why is it that only polymorphic markers can be used for genetic mapping?
3. What are sequence tagged sites (STS)? Describe examples of commonly used STS.
4. What is the rationale of developing both genetic and physical maps?
5. What is a "mapping reagent"? Describe the two common mapping reagents used.
6. Why is BAC preferred over YAC in making clone libraries?

7. What type of vectors is used in hierarchical shotgun sequencing? Explain how and why these vectors are used.
8. The BAC vector is used for making libraries as mapping reagents in physical mapping (Sections 23.2.3 and 23.2.4). It is also used in the sequencing phase (Section 23.4.1). How are these two uses connected? Can the BAC clones developed for physical mapping be used for sequencing the genome?
9. For those who are interested in learning more about the subject of human genome sequencing: visite http://www.genome.gov.

GENERAL REFERENCES

Gibson, G., and Muse, S. V. 2004. *A Primer of Genome Science.* 2nd edition, Sinauer Associates, Inc. Publishers. MA.

Kreuzer, H., and Massey, A. 2001. *Recombinant DNA and Biotechnology: A Guide for Students.* 2nd edition, American Society Microbiology, Washington, D.C.

Lewin, B. 2003. *Genes VIII.* Prentice Hall, Inc. NJ.

Primrose, S. B., Twyman, R. M., and Old, R. W. 2002. *Principles of Gene Manipulation.* 6th edition, Blackwell Publishers, London.

Ptashne, M. 2004. *A Genetic Switch.* 3rd edition, Cold Spring Harbor Press, NY.

Sambrook, J., and Russell, D. W. 2001. *Molecular Cloning. A Laboratory Manual.* 3rd edition, Cold Spring Harbor Laboratory Press, Plainview, NY.

White, R. J. 2001. *Gene Transcription: Mechanisms and Control.* Blackwell Science Ltd., Oxford.

SUGGESTED READINGS

PART ONE Fundamentals of Genetic Processes

Chapter 1: Introductory Concepts

deDuve, C. 1996. The birth of complex cells. *Sci. Am.* 274(4), 50-57.

Giddings, G. 2001. Transgenic plants as protein factories. *Curr. Opin. Biotechnol.* 12, 450-454.

Goff, S. A., and Salmeron, J. M. 2004. Back to the future of cereals. *Sci. Am.* 291(2), 42-49.

Holliday, R. 1989. A different kind of inheritance. *Sci. Am.* 260(6), 60-73.

Klug, A. 1981. The nucleosome. *Sci. Am.* 244(2), 52-64.

Langridge, W. H. R. Edible vaccines. *Sci. Am.* 283, 66-71.

Murray, A. W., and Kirschner, M. W. 1991. What controls the cell cycle. *Sci. Am.* 264(3), 56-63.

Murray, A. W., and Szostak, J. W. 1987. Artificial chromosomes. *Sci. Am.* 257(5), 62-68.

Sapienza, C. 1990. Parental imprinting of genes. *Sci. Am.* 263(4), 52-60.

Tarner, I. H., Muller-Ladner, U., and Fathman, C. G. 2004. Targeted gene therapy: frontiers in the development of 'smart drugs'. *Trends Biotechnol.* 22, 304-310

Twyman, R. M., Stoger, E., Schillberg, S., Christou, P., and Fischer, R. 2003. Molecular farming in plants: host systems and expression technology. *Trends Biotechnol.* 21, 570-578.

Chapter 2: Structures of Nucleic Acids

Darnell, J. E. Jr. 1985. RNA. *Sci. Am.* 253(4), 68-78.

Felsenfeld, G. 1985. DNA. *Sci. Am.* 253(4), 58-67.

Grivell, L. A. 1983. Mitochrondrial *DNA. Sci Am.* 248(3), 78-89.

Wallace, D. C. 1997. Mitochondrial DNA in aging and disease. *Sci.. Am.* 277(2), 40-47.

Chapter 3: Structures of Proteins

Richards, F. M. 1991. The protein folding problem. *Sci. Am.* 264(1), 54-63.

Unwin, N., and Henderson, R. 1984. The structure of proteins in biological membranes. *Sci. Am.* 250(2), 78-94.

Weinberg, R. A. 1985. The molecules of life. *Sci. Am.* 253(4), 48-57.

Chapter 4: The Genetic Process

Brosius, J. 2001. tRNAs in the spotlight during protein biosynthesis. *Trends Biochem. Sci.* 26, 653-656

Dickerson, R. E. 1983. The DNA helix and how it is read. *Sci. Am.* 249(6), 94-111.

Ernst, J. F. 1988. Codon usage and gene expression. *Trends Biotechnol.* 6(8), 196-199.

Lake, J. A. 1981. The ribosome. *Sci. Am.* 245(2), 84-97.

Moore, P. B., and Steitz, T. A. 2005. The ribosome revealed. *Trends Biochem. Sci.* 30, 281-283

Nirenberg, M. 2004. Historical review: Deciphering the genetic code- a personal account. *Trends Biochem. Sci.* 29, 47.

Nomura, M. 1984. The control of ribosome synthesis. *Sci. Am.* 250(1), 102-114.

Radman, M., and Wagner, R. 1988. The high fidelity of DNA duplication. *Sci. Am.* 259(2), 40-46.

Chapter 5: Organization of Genes

Ast, G. 2005. The alternative genome. *Sci. Am.* 292(4), 40-47.

Beardsley, T. 1991. Smart genes. *Sci. Am.* 265(2), 86-95.

Darnell, J. E. Jr. 1983. The processing of RNA. *Sci. Am.* 249(4), 90-100.

Diller, J. D., and Raghuraman, M. K. 1994. Eukaryotic replication origins: Control in space and time. *Trends Biochem. Sci.* 19, 320-325.

Gibbs, W. 2003. Unseen genome: gems among the junk. *Sci. Am.* 289(6), 48-53.

Grunstein, M. 1992. Histones as regulators of genes. *Sci. Am.* 267(4), 68-74B.

Latchman, D. S. 2001. Transcription factors: bound to activate or repress. *Trends Biochem. Sci.* 26, 211-213.

Le Hir, H., Nott, A., and Moore, M. J. 2003. How introns influence and enhance eukaryotic gene expression. *Trends Biochem. Sci.* 28, 215-220.

Kornberg, R. D. 2005. Mediator and the mechanism of transcriptional activation. *Trends Biotechnol.* 30, 235-239.

McKnight, S. L. 1991. Molecular zippers in gene regulation. *Sci. Am.* 264(4), 54-64.

Ptashne, M. 1989. How gene activators work. *Sci. Am.* 260(1), 41-47.

Ptashne, M. 2005. Regulation of transcription: from lambda to eukaryotes. *Trends Biochem. Sci.* 30, 275-279

Ptashne, M., Johnson, A. D., and Pabo, C. O. 1982. A genetic switch in a bacterial virus. *Sci. Am.* 247(5), 128-140.

Sharp, P. A. 2005. The discovery of split genes and RNA splicing. *Trends Biochem. Sci.* 30, 279-281.

Steitz, J. A. 1988. Snurps. *Sci. Am.* 258(6), 56-63.

Tjian, R. 1995. Molecular machines that control genes. *Sci. Am.* 272(2), 54-61.

Chapter 6. Reading the Nucleotide Sequence of a Gene

Fisher, L. W., Heegaard, A.-M., Vetter, U., Vogel, W., Just, W., Termine, J. D., and Young, M. F. 1991. Human biglycan gene. *J. Biol. Chem.* 266, 14371-14377.

Lundberg, L. G., Thoresson, H.-O., Karlstrom, O. H., and Nyman, P. O. 1983. Nucleotide sequence of the structural gene for dUTPase of *Escherichia coli* K-12. *EMBO J.* 2, 967-971.

Ungefroren, H., and Krull, N. B. 1996. Transcriptional regulation of the human biglycan gene. *J. Biol. Chem.* 271, 15787-15795.

PART TWO Techniques and Strategies of Gene Cloning

Chapter 7: Enzymes Used in Cloning

Bickle, T. A., and Kruger, D. H. 1993. Biology of DNA restriction. *Microbiol. Rev.* 57, 434-450.

Pavlov, A. R., Pavlova, N. V., Kozyavkin, S. a., and Slesarev, A. L. 2004. Recent developments in the optimization of thermostable DNA polymerases for efficient applications. *Trends Biotechnol.* 22, 253-260.

Roberts, R. J., and Macelis, D. 1993. REBATE-restriction enzymes and methylases. *Nucl. Acids Res.* 21, 3125-3137.

Chapter 8: Techniques Used in Cloning

Cohen, S. N. 1975. The manipulation of genes. *Sci. Am.* 233(1), 25-33.

Martin, C., Bresnick, L., Juo, R.-R., Voyta, J. C., and Bronstein, I. 1991. Improved chemiluminescent DNA sequencing. *BioTechniques* 11, 110-114.

Mullis, K. B. 1990. The unusual origin of the polymerase chain reaction. *Sci. Am.* 262(4), 56-65.

Sanger, F. 1981. Determination of nucleotide sequences in DNA. *Bioscience Reports* 1, 3-18.

Southern, E. M. 1975. Detection of specific sequences among DNA fragments separated by gel electrophoresis. *J. Mol. Biol.* 98, 503-517.

Thomas, P. S. 1980. Hybridization of denatured RNA and small DNA fragments transferred to nitrocellulose. *Proc. Natl. Acad. Sci. USA* 22, 5201-5205.

Chapter 9: Cloning vectors for Introducing Genes into Host Cells

Cameron, I. R., Possee, R. D., and Bishop, D. H. L. 1989. Insect cell culture technology in baculovirus expression system. *Trends Biotechnol.* 7, 66-70.

Chauthaiwale, V. M., Therwath, A., and Deshpande, V. V. 1992. Bacteriophage lamba as a cloning vector. *Microbiol. Rev.* 56, 577-591.

Davies, A. H. 1994. Current methods for manipulating baculoviruses. *Bio/Technology* 12, 47-50.

Hohn, B., and Collins, J. 1988. Ten years of cosmids. *Trends Biotechnol.* 6, 293-298.

Katzen, F., Chang, G., Kudlicki, W. 2005. The past, present and future of cell-free protein synthesis. *Trends Biotechnol.* 23, 150-156.

Lu, Q. 2005. Seamless cloning and gene fusion. *Trends Biotechnol.* 23, 199-207.

Luque, T., and O'Reilly, D. R. 1999. Generation of baculovirus expression vectors. *Mol. Biotechnol.* 11, 153-163.

Newell, C. A. 2000. Plant transformation technology. *Mol. Biotechnol.* 16, 53-65.

Ramsay, M. 1994. Yeast artificial chromosome cloning. *Mol. Biotechnol.* 1, 181-201.

Schenborn, E., and Groskreutz, D. 1999. Reporter gene vectors and assays. *Mol. Biotechnol.* 13, 29-43.

Schuermann, D., Molinier, J., Fritsch, O., and Hohn, B. 2005. The dual nature of homologous recombination in plants. *Trends Genetics* 21, 173-181

Simons, K., Garoff, H., and Helenius, A. 1982. How an animal virus gets into and out of its host cell. *Sci. Am.* 246(2), 58-66.

Varmus, H. 1987. Reverse transcription. *Sci. Am.* 257(3), 56-64.

Chapter 10: Transformation

Chassy, B. M., Mercenier, A., and Flickinger, J. 1988. Transformation of bacteria by electroporation. *Trends Biotechnol.* 6, 303-309.

Hockney, R. C. 1994. Recent developments in heterologous protein production in *Escherichia coli*. *Trends Biotechnol.* 12, 456-463.

Hohn, B., Levy, A. A., Puchta, H. 2001. Elimination of selection markers from transgenic plants. *Curr. Opin. Biotechnol.* 12, 139-143.

Klein, T. M., Arentzen, R., Lewis, P. A., and Fitzpatrick-McElligott, S. 1992. Transformation of microbes, plants and animals by particle bombardment. *Bio/Technology* 10, 286-290.

Maheshwari, N., Rajyalakshmi, K., Baweja, K., Dhir, S. K., Chowdhry, C. N., and Maheshwari, S. C. 1995. *In vitro* culture of wheat and genetic transformation - retrospect and prospect. *Crit. Rev. Plant Sci.* 14, 149-178.

Walden, R., and Wingender, R. 1995. Gene-transfer and plant-regeneration techniques. *Trends Biotechnol.* 13, 324-331.

Whitelam, G. C., Cockburn, B., Gandecha, A. R., and Owen, M. R. L. 1993. Heterologous protein production in transgenic plants. *Biotechnol. Genet. Engineer. Rev.* 11, 1-29.

Chapter 11: Isolating Genes for Cloning

Cohen, S. N 1975. The manipulation of genes. *Sci. Am.* 233(1), 25-33.

Kimmel, A. R. 1987. Selection of clones from libraries: Overview. *Methods Enzymol.* 152, 393-399.

Okayama, H., Kawaichi, M., Brownstein, M., Lee, F., Yokota, T., and Arai, K. 1987. High-efficiency cloning of full-length cDNA; Construction and screening of cDNA expression libraries for mammalian cells. *Methods Enzymol.* 154, 3-28.

PART THREE Impact of Gene Cloning - Applications in Agriculture

Chapter 12: Improving Tomatoe Quality by Antisense RNA

Kramer, M., Sanders, R. A., Sheehy, R. E., Melis, M., Kuehn, M., and Hiatt, W. R. 1990. Field evaluation of tomatoes with reduced polygalacturonase by antisense RNA. In: *Horticultural Biotechnology*, eds. A. B. Bennett and S. D. O'Neil, Wiley-Liss Inc., New York.

Lau, N. C., and Bartel, D. P. 2003. Censors of the genome. *Sci. Am.* 289(2), 34-41.

Schuch, W. 1994. Improving tomato quality through biotechnology. *Food Technology* 48(11), 78-83.

Sheehy, R. E., Kramer, M., and Hiatt, W. R. 1988. Reduction of polygalacturonase activity in tomato fruit by antisense RNA. *Proc. Natl. Acad. Sci. USA* 85, 8805-8809.

Sheehy, R. E., Pearson, J., Brady, C. J., and Hiatt, W. R. 1987. Molecular characterization of tomato fruit polygalacturonase. *Mol. Gen. Genet.* 208, 30-36.

Wagner, R. W. 1994. Gene inhibition using antisense oligodeoxynucleotides. *Nature* 372, 333-335.

Chapter 13. Transgenic Crops Engineered with Insecticidal Activity

Ffrench-Constant, R. H., Daborn, P. J., and Le Goff, G. The genetics and genomics of insecticide resistance. 2004. *Trends Genetics* 20, 164-170.

Fischhoff, D. A., Bowdish, K. S., Perlak, F. J., Marrone, P. G., McCormick, S. M., Niedermeyer, J. G., Dean, D. A., Kusano-Kretzmer, K., Mayer, E. J., Rochester, D. E., Rogers, S. G., and Fraley, R. T. 1987. Insect tolerant transgenic tomato plants. *Bio/Technology* 5, 807-813.

Koziel, M. G., Carozzi, N. B., Currier, T. C., Warren, G. W., and Evola, S. V. 1993. The insecticidal crystal proteins of *Bacillus thuringiensis*: Past, present and future uses. *Biotechnol. Genet. Engineer.* Rev. 11, 171-228.

Perlak, F. J., Deaton, R. W., Armstrong, T. A., Fuchs, R. L., Sims, S. R., Greenplate, J. T., and Fischhoff, D. A. 1990. Insect resistant cotton plants. *Bio/Technology* 8, 939-943.

Rietschel, E. T., and Brade, H. 1992. Bacterial endotoxins. *Sci. Am.* 267(2), 54-61.

Shah, D. M., Rommems, C. M. T., and Beachy, R. N. 1995. Resistance to diseases and insects in transgenic plants: progress and applications to agriculture. *Trends Biotechnol.* 13, 362-368.

Umbeck, P., Johnson, G., Barton, K., and Swain, W. 1987. Genetically transformed cotton (*Grossypium hirsutum* L.) plants. *Bio/Technology* 5, 263-266.

Zhao, J.-Z., Cao, J., Li, Y., Collins, H. I., Roush, R. T., Earle, E. D., and Shelton, A. M. 2003. Transgenic plants expressing two *Bacillus thuringiensis* toxins delay insect resistance evolution. *Nature Biotechnology* 21, 1493-1497.

Chapter 14: Transgenic Crops Conferred with Herbicide Resistance

Comai, L., Facciotti, D., Hiatt, W. R., Thompson, G., Rose, R. E., and Stalker, D. M. 1985. Expression in plants of a mutant *aroA* gene from *Salmonella typhimurium* confers tolerance to glyphosate. *Nature* 317, 741-744.

Gurr, S. J., and rushton, P. J. 2005. Engineering plants with increased disease resistance: how are we going to express it? *Trends Biotechnol.* 23, 283-290.

Hines, P. J., and Marx, J. (eds.) 1995. The emerging world of plant science. *Science* 268, 653-716.

Fillatti, J. J., Kiser, J., Rose, R., and Comai, L. 1987. Efficient transfer of a glyphosate tolerance gene into tomato using a binary *Agrobacterium tumefaciens* vector. *Bio/Technology* 5, 726-730.

McDowell, J. M., Woffenden, B. J. 2003. Plant disease resistance gene: recent insights and potential applications. *Trends Biotechnol.* 21, 178-183.

Schulz, A., Wengenmayer, F., and Goodman, H. M. 1990. Genetic engineering of herbicide resistance in high plants. *Plant Science* 9, 1-15.
Strobel, G. A. 1991. Biological control of weeds. *Sci. Am.* 265(1), 72-78.

Chapter 15: Growth Enhancement in Trangenic Fish

Chen, T. T., Lin, C.-M., Lu, J. K., Shamblott, M., and Kight, K. 1993. Transgenic fish: a new emerging technology for fish production. In: *Science for The Food Industry of The 21st Century, Biotechnology, Supercritical Fluids, Membrances and Other Advanced Technologies for Low Calorie, Healthy Food Alternatives.* ed. M. Yalpani, ATL Press, Mount Prospect, IL.
Chen, T. T., and Powers, D. A. 1990. Transgenic fish. *TIBTECH* 8, 209-215.
Du, S. J., Gong, Z., Fletcher, G. L., Shears, M. A., King, M. J., Idler, D. R., and Hew, C. L. 1992. Growth enhancement in transgenic Atlantic salmon by the use of an "all fish" chimeric growth hormone gene construct. *Bio/Technology* 10, 176-181.

PART FOUR Impact of Gene Cloning - Applications in Medicine

Chapter 16: Microbial Production of Recombinant Human Insulin

Atkinson, M. A., and Maclaren, N. K. 1990. What causes diabetes? *Sci. Am.* 263(1), 62-71.
Bristow, A. F. 1993. Recombinant-DNA-derived insulin analogues as potentially useful therapeutic agents. *Trends Biotechnol.* 11, 301-305.
Gilbert, W., and Willa-Komaroff, L. 1980. Useful proteins from recombinant bacteria. *Sci. Am.* 242(4), 74-94.
Goeddel, D. V., Kleid, D. G., Bolivar, F., Heyneker, H. L., Yansura, D. G., Crea, R., Hirose, T., Kraszewski, A., Itakura, K., and Riggs, A. D. 1979. Expression in *Escherichia coli* of chemically synthesized genes for human insulin. *Proc. Natl. Acad. Sci. USA* 76, 106-110.
Lienhard, G. E., Slot, J. S., James, D. E., and Mueckler, M. M. 1992. How cells absorb glucose? *Sci. Am.* 266(1), 86-91.

Chapter 17: Finding Disease-Causing Genes

Anand, R. 1992. Yeast artificial chromosomes (YACs) and the analysis of complex genomes. *Trends Biotechnol.* 10, 35-40.

Buckler, A. J., Chang, D. D., Graw, S. L., Brook, D., Haber, D. A., Sharp, P. A., and Housman, D. E. 1991. Exon amplification: a strategy to isolate mammalian genes based on RNA splicing. *Proc. Natl. Acad. Sci. USA* 88, 4005-4009.

Collins, F. S. 1991. Of needles and haystacks: Finding human disease genes by positional cloning. *Clin. Res.* 39, 615-623.

Halaas, J. L., Gajiwala, K. S., Maffei, M., Cohen, S. L., Chaiti, B. T., Rabinowitz, D., Lallone, R. L., Burley, S. K., Friedman, J. M. 1995. Weight-reducing effects of the plasma protein encoded by the obese gene. *Science* 269, 543-544.

Pelleymounter, M. A., Cullen, M. J., Baker, M. B., Hecht, R., Winters, D., Boone, T., and Collins, F. 1995. Effects of the obese gene product on body weight regulation in *ob/ob* mice. *Science* 269, 540-543.

Poustka, A., Pohl, T. M., Barlow, D. P., Frischauf, A.-M., and Lehrach, H. 1987. Construction and use of human chromosome jumping libraries from *Not*I-digested DNA. *Nature* 325, 353-355.

White, R., and Lalouel, J.-M. 1988. Chromosome mapping with DNA markers. *Sci. Am.* 258(2), 40-48.

Zhang, Y., Proenca, R., Maffei, M., Barone, M., Leopold, L., and Friedman, J. M. 1994. Positional cloning of the mouse obese gene and its human homologue. *Nature* 372, 425-432.

Chapter 18: Human Gene Therapy

Blaese, R. M. 1997. Gene therapy for cancer. Sci. Am. 276, 111-120.

Blaese, R. M., Culver, K. W., Miller, A. D., Carter, C. S., Fleisher, T., Clerici, M., Shearer, G., Chang, L., Chiang, Y., Tolstoshew, P., Greenblatt, J. L., Rosenberg, S. A., Klein, H. Berger, M., Mullen, C. A., Ramsey, J., Muul, L., Morgan, R. A., and Anderson, W. F. 1995. T lymphocyte-directed gene therapy for ADA-SCID: Initial trial results after 4 years. *Science* 270, 475-480.

Felgner, P. L. 1997. Nonviral strategies for gene therapy. *Sci. Am.* 276(6), 102-106.

Hock, R. A., Miller, D., and Osborne, W. R. A. 1989. Expression of human adenosine deaminase from various strong promoters after gene transfer into human hematopoietic cell lines. *Blood* 74, 876-881.

Lundstrom, K. 2003. Latest development in viral vectors for gene therapy. *Trends Biotechnol.* 21, 117-122.

Morgan, R. A., and Anderson, W. F. 1993. Human gene therapy. *Ann. Rev. Biochem.* 62, 191-217.

Mountain, A. 2000. Gene therapy: the first decade. *Trends Biotechnol.* 18, 119-128.

Sheikh, N. A., and Morrow, W. J. W. 2003. Guns, genes, and spleen: a coming of age for rational vaccine design. *Methods* 31, 183-192.

Verma, I. M. 1990. Gene therapy. *Sci. Am.* 263(5), 68-84.

Weiner, D. B., and Kennedy, R. C. 1999. Genetic vaccines. *Sci. Am.* 281(1), 50-57.

Chapter 19: Gene Targeting

Capecchi, M. R. 1994. Targeted gene replacement. *Sci. Am.* 270(3), 52-59.

Capecchi, M. R. 2000. How close are we to implementing gene targeting in animals other than the mouse. *Proc. Natl. Acad. Sci. USA* 97, 956-957.

Clark, A. J., Burl, S., Denning, C., and Dickinson, P. 2000. Gene targeting in livestock: a preview. *Transgenic Res.* 9, 263-275.

Dai, Y., Vaught, T. D., Boone, J., Chen, S.-H., Phelps, C. J., Ball, S., Monahan, J. A., Jobst, P. M., McCreath, K. J., Lamborn, A. E., Cowell-Lucero, J. L., Wells, K. D., Colman, A., Poejaeva, I. A., and Ayares, D. L. 2002. Targeted disruption of the α-1,3-galactosyltransferase gene in cloned pigs. *Nature Biotechnol.* 20, 251-255.

Fassler, R., Martin, K., Forsberg, E., Litzenburger, T., and Iglesias, A. 1995. Knockout mice: How to make them and why. The immunological approach. *Int. Arch. Allergy Immunol.* 106, 323-334.

Kim, H.-S., and Smithies, O. 1988. Recombinant fragment assay for gene targeting based on the polymerase chain reaction. *Nucleic Acids Res.* 16, 8887-8903.

Kolber-Simonds, D., Lai, L., Watt, S. R., Denaro, M., Arn, S. Augenstein, M. L., Betthauser, J., Carter, D. B., Greenstein, J. L., Hao, Y., Im, G.-S., Liu, Z., Mell, G. D., Murphy, C. N., Park, K.-W., Rieke, A., Ryan, D. J. J., Sachs, D. H., Forsberg, E. J., Prather, R. S., and Hawley, R. J. 2004. Production of α-1,3-galactosyltransferase null pigs by means of nuclear transfer with fibroblasts bearing loss of heterozygosity mutations. *Proc. Natl. Acad. Sci.* 101, 7335-7340.

Lai, L., and Prather, R. S. 2003. Creating genetically modified pigs by using nuclear transfer. *Repro. Biol. Endocrinol.* 1, 82-87.

Lanza, R. P., Cooper, D. K. C., and Chick, W. L. 1997. Xenotransplantation. *Sci. Am.* 277, 54-59.

Smith, K. R. 2002. Gene transfer in higher animals: theoretical considerations and key concepts. *J. Biotechnol.* 99, 1-22.

Chapter 20: DNA Typing

Carey, L., and Mitnik, L. 2002. Trends in DNA forensic analysis. *Electrophoresis* 23, 1386-1397.

Debenham, P. G. 1992. Probing identity: The changing face of DNA fingerprinting. *Trends Biotechnol.* 10, 96-102.

Jeffreys, A. J., Turner, M., and Debenham, P. 1991. The efficiency of multi-locus DNA fingerprint probes for individualization and establishment of family relationships, determined from extensive casework. *Am. J. Hum. Genet.* 48, 824-840.

Gill, P. 2002. Role of short tandem repeat DNA in forensic casework in the UK – past, present, and future perspectives. *BioTechniques* 32, 366-385.

Moxon, E. R., and Wills, C. 1999. DNA microsatellites: agents of evolution? *Sci. Am.* 280(1), 94-99.

Nakamura, Y., Leppert, M., O'Connell, P., Wolff, R., Holm, T., Culver, M., Martin, C., Fujimoto, E., Hoff, M., Kumlin, E., and White, R. 1987. Variable number of tandem repeat (VNTR) markers for human gene mapping. *Science* 235, 1616-1622.

Ruitberg, C. M., Reeder, D. J., and Butler, J. M. 2001. STRBase: a short tandem repeat DNA database for the human identity testing community. *Nucleic Acids Res.* 29, 320-322.

Chapter 21: Transpharmers - Bioreactors for Pharmceutical Products

Bawden, W. S., Passey, R. T., and Mackinlay, A. G. 1994. The genes encoding the major milk-specific protein and their use in transgenic studies and protein engineering. *Biotechnol. Genet. Engineer. Rev.* 12, 89-137.

Carver, A. S., Dalrymple, M. A., Wright, G., Cotton, D. S., Reeves, D. B., Gibson, Y. H., Keenan, J. L., Barrass, J. D., Scott, A. R., Colman, A., and Garner, I. 1993. Transgenic livestock as bioreactors: Stable expression of human alpha-1-antitrypsin by a flock of sheep. *Bio/Technology* 11, 1263-1270.

Fischer, R., and Emans, N. 2000. Molecular farming of pharmaceutical proteins. *Transgenic Res.* 9, 279-299.

Ivarie, R. 2003. Avian transgenesis: progress towards the promise. *TIBTECH* 21, 14-19.

Maga, E. A., and Murray, J. D. 1995. Mammary gland expression of transgenes and the potential for altering the properties of milk. *Bio/Technology* 13, 1452-1457.

Wright, g., Carver, A., Cottom, d., Reeves, D., Scott, A., Simons, P., Wilmut, I., Garner, I., and Colman, A. 1991. High level expression of active human alpha-1-antitrypsin in the milk of transgenic sheep. *Bio/Technology* 9, 830-834.

Chapter 22: Animal Cloning

Lanza, R. P., Dresser, B. L., and Damiani, P. 2000. Cloning Noah's Ark. *Sci. Am.* 283(5), 84-89.

Matthew, J. E., Gutter, C., Loike, J. D., Wilmut, I., Schnieke, A. E., and Schon, E. A. 1999. Mitochondrial DNA genotypes in nuclear transfer-derived cloned sheep. *Nature Genetics* 23, 90-93.

Roslin Institute. 2000. Nuclear transfer: a brief history. www.roslin.ac.uk/public/01-03-98-nt.html.

Wilmut, I. 1998. Cloning for medicine. *Sci. Am.* 279(6), 58-63.

Wilmut, I., Beaujean, N., de Sousa, P. A., Dinnyes, A., King, T. J., Paterson, L. A., Wells, D. N., and Young, L. E. 2002. Somatic cell nuclear transfer. *Nature* 419, 583-586.

Chapter 23: Human Genome Sequencing

Bentley, D. R., Fruitt, K. D., Deloukas, P., Schuler, G. D., and Ostell, J. 1998. Coordination of human genome sequencing via a consensus framework map. *Trends Genetics* 14, 381-384.

Deloukas, P., et al. 1999. A physical map of 30,000 human genes. *Science* 282, 744-746.

Dib, C., et al. 1996. A comprehensive genetic map of the human genome based on 5,264 microsatellites. *Nature* 380, 152-154.

Hudson, T. J., et al. 1995. A STS-based map of the human genome. *Science* 270, 1945-1954.

International Human Genome Sequencing Consortium. 2001. Initial sequencing and analysis of the human genome. *Nature* 409, 860-921.

March, R. E. 1999. Gene mapping by linkage and association analysis. *Mol. Biotechnol.* 13, 113-122.

Murray, J. C., et al. 1994. A comprehensive human linkage map with centimorgan density. *Science* 265, 2049-2054.

Nachman, M. W. 2001. Single nucleotide polymorphisms and recombination rate in humans. *Trends Genetics* 17, 481-485.

Oliver, M., et al. 2001. A high-resolution radiation hybrid map of the human genome draft sequence. *Science* 291, 1298-1302.

Shizuya, H., Birren, B., Kim, U.-J., Mancino, V., Slepak, T., Tachiri, Y., and Simon, M. 1992. Cloning and stable maintenance of 300-kilobase-pair fragments of human DNA in *Escherichia coli* using an F-factor-based vector. *Proc. Natl. Acad. Sci. USA* 89, 8794-8797.

Shizuya, H., and Kouros-Mehr, H. 2001. The development and applications of the bacterial artificial chromosome cloning system. *Keio J. Med.* 50, 26-30.

Venter, J. C., Smith, H. O., and Hood, L. 1996. A new strategy for genome sequencing. *Nature* 381, 364-366/

Venter, J. C., et al. 2001. The sequence of the human genome. *Science* 291, 1304-1351.

White, R., and Lalouel, J.-M. 1988. Chromosome mapping with DNA markers. *Sci. Am.* 258(2), 40-48.

INDEX

J

K

L

M

N

O

P

U

V

W

X

Y

Z